Concepts in Chemistry

Crystals and Crystal Structure

Concepts in Chemistry

Titles in this series:

Crystals and Crystal Structure

Michael Hudson BSc ARIC PhD
St Paul's School, Barnes, London S.W.13

Longman

LONGMAN GROUP LIMITED
London
Associated companies, branches and
representatives throughout the world

© Michael Hudson 1971

First published 1971
ISBN 0 582 32139 5

Printed in Great Britain by
William Clowes & Sons, Limited
London, Beccles and Colchester

Contents

Introduction

This book on crystals and crystallography has been designed to follow a wide range of syllabuses and is suitable for students of physics, chemistry, geology, biochemistry, metallurgy and crystallography. Students both at school and at university will find material that will help their studies. Readers at school will find Chapters 1, 2, 4, 5 and 6 the most useful and Chapters 4 and 5 are designed so that they can refer quickly to the correct structure. A special effort has been made to provide a pictorial approach and figures are closely integrated with the text. In addition, certain experiments have been suggested and references and questions are at the ends of the chapters. Examples have been carefully chosen with one of three objects in mind. First, some structures such as the fluorite structure have been included, since they emphasise important aspects of simple structures. Secondly, structures such as hexamethylbenzene have been included because they are historically important. Thirdly, a knowledge of the structures of some biochemical molecules such as lysozyme is helping scientists to understand the nature of life itself. The construction of models should help students to appreciate crystal structures and the use of the 0·03 metre waves will enable students to see the important aspects of X-ray crystallography. The reader may wish to verify some of the statements made by building his own models and by experimentation. (The Nuffield O-level Chemistry Project *Handbook for Teachers* deals thoroughly with the construction of models.) A major problem in writing about crystals is not what to include but what to leave out; it is hoped that sufficient basic material has been included in this book to guide the student toward further studies.

Michael Hudson
1970

Acknowledgements

I am particularly grateful to my father who typed the original manuscript, and wish to thank the following for their help and for permission to reproduce illustrations:

Sir Lawrence Bragg made some useful suggestions for the early chapters of the book; Dr C. W. Bunn and Oxford University Press for Fig. 3.18 from *Chemical Crystallography*, 2nd ed. 1961; Dr E. C. Cocking of the Department of Botany at the University of Nottingham for Plates 2.11, 2.12 and 7.9; Dr Dorothy Crowfoot-Hodgkin for Figs 7.15 and 7.16; Crystal Structures Ltd, Bottisham, Cambridge for Plate 1.4 and for allowing my brother and me to photograph their models which can be seen in Plates 5.1, 5.2, 6.1, 6.2, 6.3, 6.5 and 7.1; Professor C. E. Dent of the University College Hospital for Plate 1.3; G.E.C. and A.E.I. (Electronics) Limited, Scientific Apparatus Division, for Plates 2.8 and 2.9; Dr R. W. Home of the Agricultural Research Council for Plate 7.8; Dr J. A. Kitchener of the Department of Mining and Mineral Technology at the Imperial College of Science and Technology for Plates 2.7 and 2.10; Dame Kathleen Lonsdale for Plates 2.4 and 3.2 and Fig. 6.7 (*Chemistry and Industry*, 1966) and Figs 6.8 and 7.20, and for grateful advice regarding several chapters; Mullard Ltd, for Plate 2.2; Mr W. T. Nichols of the Ministry of Technology for Plates 6.7 and 6.8; Oxford University Press Ltd, for Fig. 5.12 from A. F. Wells, *Structural Inorganic Chemistry*, 3rd ed. 1962; Science Museum, London for Plates 1.1, 1.2, 2.1, 2.3 and 7.11 (Crown Copyright reserved) and Plate 2.13 which is lent to the Science Museum by Peter Spence & Sons Ltd, Farnworth, Lancs; Plates 2.5, 2.6, 6.6 are Shell photographs; Dr M. F. Perutz used Plate 7.2 and Fig. 7.8 in his paper on 'The X-ray Analysis of Haemoglobin' which he gave on receiving the Nobel Prize. Similarly Dr J. C. Kendrew provided Plates 7.3 and 7.4 from his papers on Myoglobin; Professor D. C. Phillips for Plate 7.5 (photograph by Winifred Browne, Laboratory of Molecular Biophysics, Oxford); Pye Unicam Ltd for Plate 3.1; Professor R. E. Smallman of the University of Birmingham for Plate 6.4; Dr R. J. Valentine of the Medical Research Council for reading certain sections of the book and for Fig. 7.11 and Plates 7.6, 7.7 and 7.10; Dr P. F. Weller for Fig. 6.12(a) from *J. Chemical Education*, 1967, **44**, 391; Professor Waser for Fig. 3.25 from *J. Chemical Education*, 1968, **48**, 46; Dr E. A. D. White, now of the Imperial College of Science and Technology, for Figs 1.6, 1.7, 1.9, 1.10 and 1.11 first used in 'Crystal Growth Techniques', *G.E.C. Journal*, Volume 31, 1964, Figs 1.7 and 1.14 which appeared in 'The Synthesis and Use of Artificial Gemstones', *Endeavour*, April 1962; John Wiley & Sons, Australasia Pty. Ltd, for Fig. 4.21 from Aylward and Findlay, *The Chemical Data Book*.

I am also indebted to the following people who read certain sections of the book: Dr R. E. Yorke, Mr M. Segal, Mr R. B. Morris and Mr D. Pirkis.

I am extremely grateful for the advice and help that the following people gave me at the galley proof stage: Dr E. A. D. White, Dr C. J. Brown, Mr P. E. D. Hughes, Dr B. C. Stace, Mr J. M. Towner, Mr P. Lascalles, Dr C. John Booker.

M. H.

1 The uses and preparation of crystals

First, it is important to answer the question 'Why study crystals?' This question can be most easily answered by describing briefly how they were used in the past and how they are used today. The Assyrians and Egyptians of old used crystals as means of decoration, and they have been used for flavouring, for cattle lick, for scouring, for lubrication, for mordants and for tools. Indeed, one of the earliest uses of crystals was probably as tools and weapons. The value of some was increased by their scarcity and diamonds, sapphires, rubies, spinels, emeralds, corundum and garnets were recognised as precious stones. Even today we recognise the importance of these gems, but we now have available to us a very wide range of new synthetic crystals which, in some cases, are more pure and hard-wearing, and just as beautiful as the natural ones.

Many crystals are coloured and many of the common names given to colours are synonymous with the name of the crystal. For example, crystals of jet are black and are really fossilised wood. Emeralds are green; they are a variety of beryl, which is a six-sided crystal found in granite or granitic river gravels. Calcite is found in limestone districts and may be distinguished from quartz crystals by the relative ease with which the crystals cleave. Amethysts are purple or a violet translucent colour, and are six-sided cystals found in veins cutting older rocks. Tourmaline is not quite so well known as emerald, but can occur as either black, pink, green or colourless crystals, and is often found in granites.

In Fig. 1.1 we have listed some presentday uses of the more common crystals. Tourmaline and quartz crystals were used in the Asdic apparatus during the First World War. With the advent of the Second World War supplies of sapphires and quartz from South America were greatly reduced and it was necessary to look for new ways of preparing these crystals. The preparation of crystals will be very briefly discussed.

One of the major uses of crystals is in 'windows'. For the normal visible region of radiation, glass—which is not crystalline—is readily used. In other regions of the spectrum, that is in the ultraviolet or infrared regions, corundum, fluorite or quartz may be used. One of the ways in which gramophones work is dependent on the piezo-electric effect. 'Piezo-electric' is a complicated word; it means that when a crystal of quartz is subject to a shock, for example, when it moves over the uneven edges of the groove of a gramophone record, an electric pulse is generated. This pulse is later amplified so that the listener can hear what is recorded. Alumina is sometimes used as an insulator and also in a Maser, in which vibrations in the microwave region, which is mentioned in Chapter 3, are amplified. Lasers in which coherent light or infrared radiation of high intensity is generated are discussed in Chapter 6.

Diamond and graphite are different forms of the same element carbon, and are said to be allotropes of carbon. A study of the crystal structure of diamond and graphite can answer the question, 'Why do diamond and graphite have very different physical properties if they consist only of carbon'? Diamond is one of the hardest substances known, whereas

Crystal	Doping agent	Uses
α-Al_2O_3, TiO_2, CaF_2	Transition elements	Paramagnetic studies
α-Al_2O_3, $Y_3Al_5O_{12}$	Rare earths and actinides	Masers; lasers
ZnS, CdS, organic crystals	Mn, Cu, Ag, Tl, etc.	Fluorescence Photoconductivity Photoelectricity
Ge, Si, InSb, GaAs, SiC, PbTe, Bi_2Te_3	Donor or acceptor impurities	Semiconductivity Thermoelectric Thermomagnetic Galvanomagnetic effects
Fe_3O_4, MFe_2O_4, $BaFe_{12}O_{19}$, $Y_3Fe_5O_{12}$	Paramagnetic substituents	Magnetic studies
BeO, MgO, α-Al_2O_3, UO_2	Pure	Reactor materials
Al_2SiO_5 alumino-silicates $ZrSiO_4$, C, BN, WC, ThO_2 ZrO_2, Si_3N_4, etc.	Pure	Refractories Abrasives Structural materials
Alkali halides, α-SiO_2, CaF_2, Tl(Cl, Br)	Pure	Optical materials

Fig. 1.1. The uses of crystals

graphite is a very good lubricant and may be added to oils in cars to reduce piston wear.

In Plate 1.1 a molecular model of parts of two layers of graphite is illustrated. The black balls represent the carbon atoms in the graphite structure. The structure of graphite when compared with that of diamond, which is illustrated in Plate 1.2, can be seen to have a layer framework of atoms. The carbon atoms in graphite are able to move over one another since the layers move over one another. In diamond, however, the carbon atoms are rigidly fixed, giving rise to the very hard structure of diamond.

The study of crystals is also of major importance because it gives information regarding certain illnesses. The pain in gout is caused by uric acid crystals in the joints, and silicosis is currently thought to be connected with the piezo-electric effect in crystals, which is discussed later in this book. The uric acid crystals are not particularly pointed but smooth kidney stones can be painful, and it is just possible that the very sharpness of some crystals leads to silicosis. The illness does seem to be connected with sharp crystals as it has been shown that it is not consequent on an atmosphere of smooth silica crystals. It is probable that these sharp crystals pierce or scrape the lining of the lung.

Many arthritic and rheumatoid illnesses are connected with the fact that the patient has grown crystals of varying types between the joints. If the

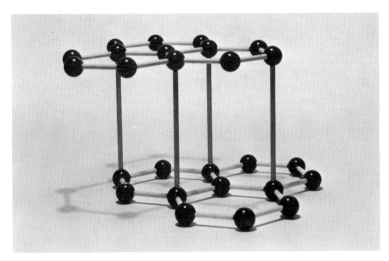

Plate 1.1. A molecular model of parts of two layers of graphite
Note that in crystallographic convention models are not usually shown in perspective

Plate 1.2. A molecular model of part of a diamond structure

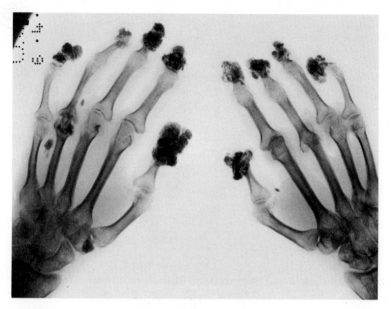

Plate 1.3. X-ray of hands
The dark patches are where crystals of apatite are grown

crystals are small they will only cause pain when the patient moves, although sometimes the crystals are large and provide pressure points giving continual pain. In Plate 1.3 the radiograph of a girl's hands is shown. A close look at the photograph shows dense masses at the ends of her fingers and thumbs and beside certain joints in her left hand. It has been shown by close work between University College Hospital and the Department of Crystallography at University College, London, that these are crystals of apatite.* The patient had a biochemical abnormality in her blood which led to the growth of colloidal-sized apatite crystallites in an organic matrix (suspending medium) which form a kind of pus where the ends of the bones should have been. The 'pus' caused swellings at the finger tips which had to be lanced to let the pus out and to relieve pain.

Recently many Nobel prizes have been awarded for some wonderful work on the evaluation of the molecular structure of many biological molecules. Some of these involved the application of X-ray studies on crystals to determine their structures. Although the subject of biological molecules is discussed in Chapter 8 it is worth while having a look at Plate 1.4 which shows the molecular structure of DNA. Comparison of this picture with

* The formula of apatite is $Ca_5(PO_4)_3F$ and indicates that the structure of apatite is built up of Ca^{2+}, PO_4^{3-} and F^- ions. The structure is complex but is isomorphous with the hydroxy-compound $Ca_5(PO_4)_3$ (OH) being similar also to olevenite (the structure of which is given in an advanced text, such as A. F. Wells, *Structural Inorganic Chemistry* 3rd edn., Oxford University Press, 1962).

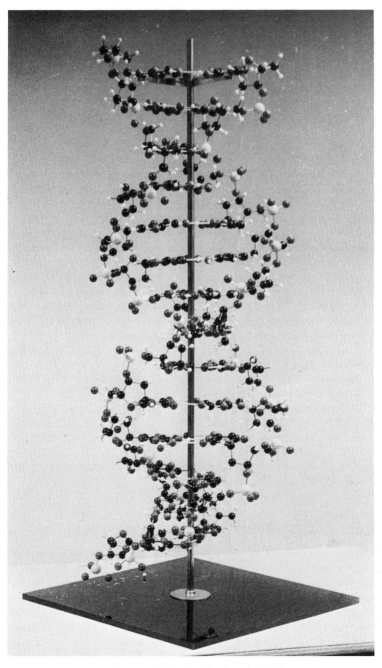

Plate 1.4. Molecular structure of DNA

that of graphite shows that it is now possible to study some very complex molecules. These molecules lead to a study of the nature of life itself, which involves the chemical and physical reactions between these complex molecules and some simpler molecules. In order to study these reactions it is vital to know the structure of the molecules concerned.

How are crystals grown?

Many of man's attempts to prepare crystals have depended on the copying of the conditions under which crystals are likely to be formed in nature. It is probable that crystals in nature were prepared under conditions which involved high temperatures and high pressures and very long times for cooling and as such it is unlikely that some crystals can be prepared under simple laboratory conditions or at home. There are, however, a number which can be prepared and these are now discussed. Most people know that many crystals are more soluble in hot water than in cold. Unfortunately it is not possible to be so vague in science and it is necessary to define what is meant by 'the solubility of a substance'. The solubility of a substance at a given temperature is defined as the mass of the substance which is dissolved by one hundred grammes of the solvent to give a saturated solution. The crystal is termed the solute and the water is referred to as the solvent; the two together are called a solution. In technical language it is said that the solute is dissolved by the solvent. The growth of crystals is a very critical process and must involve the following:
1. Small temperature variations.
2. Slow and constant crystal growth.
3. Small temperature differences within the solution or melt.
4. High purity both of solute and solvent.
5. A means of agitation or rotation of the solution or melt must be available if required.
6. The solute must be soluble at the high temperature and much less soluble at the low temperature.

Laboratory preparation of some crystals

The requirements are as above, and the solubilities of some common chemicals are listed in Fig. 1.2. The lefthand column represents the name of the substance and the chemical formula or short-hand is written in the second column. The solubility at 100 C (373 K) and that at 0 C (273 K) are then represented. Ideally there should be a large difference between the two solubilities in order that crystals are formed when the solvent cools. Thus it is a great deal easier to grow crystals of caesium nitrate than those of lead iodide by dissolving the two solutes separately in water. The solubilities may be determined by the following experiments.

Substance	Formula	Solubility in grammes per 100 grammes water at 373 K	Solubility in grammes per 100 grammes water at 273 K	Difference grammes
Sodium chloride	NaCl	39·8	35·7	4·1
Copper(II) sulphate	$CuSO_4 5H_2O$	76	14	62
Potassium nitrate	KNO_3	246	13	233
Sodium carbonate	$Na_2CO_3 10H_2O$	38·8 (at 300 K)	7	31·8
Sodium bicarbonate	$NaHCO_3$	16·4 (at 330 K)	6·9	9·5
Potassium bromide	KBr	104	54	50
Caesium nitrate	$CsNO_3$	163	9·3	153·7
Nickel(II) sulphate	$NiSO_4 7H_2O$ and $NiSO_4 . 6H_2O$	77	27	50
Cobalt(II) sulphate	$CoSO_4 6H_2O$	83	25·5	58·5
Sodium phosphate	$Na_3PO_4 . 12H_2O$	108	1·5	106·5
Alum	$(NH_4)_2SO_4$ and $Al_2(SO_4)_3 . 24H_2O$	27 (at 330 K)	2·1	24·9
Lead(II) chloride	$PbCl_2$	3·34	0·67	2·67
Lead(II) iodide	PbI_2	0·43	0·04	0·39
Calcium(II) acetate-dihydrate	$Ca(CH_3COO)_2 . 2H_2O$	32·7 (at 330 K)	37·4*	
Lithium(I) carbonate	Li_2CO_3		0·72	1·54
Lithium(I) sulphate	Li_2SO_4		23	26·1
Strontium(II) acetate-semihydrate	$Sr(CH_3COO)_2 . \frac{1}{2}H_2O$		36·4	43

* The solubility of some substances decreases with temperature.
The solubilities of gases decrease with increasing temperature.

Fig. 1.2. Solubility of different compounds

Experiment 1.1 Preparation of crystals from saturated solutions

Measure out 100 cm³ of water and place the solvent in a 200 cm³ beaker. Then take a suitable weight of the solute which is intermediate between the two extreme solubilities. Heat the solution until the solute has dissolved. Allow the solution to cool slowly: the longer the solution is allowed to cool the better the final shape of the crystal. The solution when the solute is crystallising out is said to be a saturated solution. A saturated solution is defined as the solution when the dissolved solute and the crystallised solute are in equilibrium. When a solution is saturated the solvent cannot hold any more solute.

Experiment 1.2 The solubility curve for potassium nitrate

In Fig. 1.3 the solubility curve for potassium nitrate is illustrated, and represents the solubility of potassium nitrate at different temperatures. Thus (0·246 kg) of potassium nitrate dissolves in (0·1 kg) of water at (373 K). The solubility curve may be found by the following method: Potassium nitrate is added to a beaker (250 cm³) which contains water (100 cm³). The solution is heated until the solute has dissolved and the solution allowed to cool slowly. Record the temperature at which the first precipitation is seen. Stirring is essential in this experiment because a supersaturated solution may be formed. Plot the different weights of potassium nitrate against the temperature at which precipitation is first seen.

A supersaturated solution is one in which more solute is dissolved in the solvent than would be expected from the solubility curve. Sodium thiosulphate forms a supersaturated solution in water.

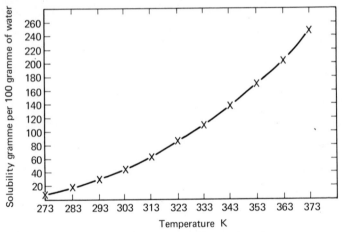

Fig. 1.3. The solubility curve of potassium nitrate

Experiment 1.3 A supersaturated solution of sodium thiosulphate

Determine the solubilities of sodium thiosulphate at 313 K (40 C) and at room temperature. Then dissolve an amount of solute which is intermediate between the two solubilities and allow the solution to cool slowly. When the solution is cooling be careful not to disturb the solution otherwise normal precipitation will occur. When the solution has been cooled add a crystal of the solute and precipitation should occur. A supersaturated solution is said to be metastable, that is, the phase which is normally stable under the given conditions does not form unless there is a small amount of the stable phase already present.

In Figs. 1.4 and 1.5 the apparatus used to prepare crystals at home or in a simple laboratory is illustrated. 'But how do crystals form?' When a

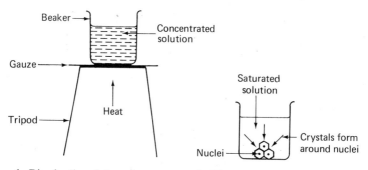

1. Dissolve the substance in water

2. When the solution cools a saturated solution is formed and crystals should start forming round a nucleus

Fig. 1.4. The method of preparing crystals

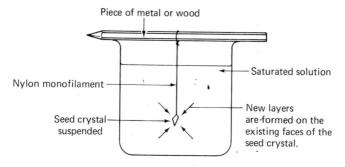

Fig. 1.5. Use of a seed crystal

For copper(II) sulphate crystals a seed crystal may be added which acts as a nucleus

solution cools the crystals start to form around a nucleus which may be another crystal of the solute or even a piece of dust. In a supersaturated solution there is no nucleus around which the crystals may form. The crystal which was added to the thiosulphate to bring about precipitation is generally called a 'seed crystal' which acts as a nucleus. In the preparation of copper sulphate a seed crystal which should be as perfect as possible, must be grown from a previous experiment.

Experiment 1.4 Copper(II) sulphate crystals

Although copper(II) sulphate is often used in schools it is not one of the easiest crystals to grow. The copper(II) sulphate is dissolved in water and the solution is allowed to cool; a seed crystal is added before precipitation occurs. The solution is left to stand for several days and a crystal forms around the nucleus or seed crystal. Can you prepare crystals of the other substances such as chrome alum (Fig. 1.2) and use these as seed crystals for growing larger crystals of the same substance? Alternatively it might be possible to carry out some of the following industrial preparations of crystals using laboratory versions of the apparatus. Certainly it should be possible to use Figs. 1.6 to 1.16 as bases of the design for laboratory apparatus and in particular zone-refining lends itself very well to the laboratory and is often used to prepare crystals for research. Unfortunately the pressures required in the manufacture of diamonds are far too high, but it may well be possible for sixth-formers under proper supervision to make a plasma flame apparatus. The steady growth of the layers of camphor could be studied (Fig. 1.19) in a school laboratory.

Industrial preparation of gems and useful crystals

There are two main methods by which crystals are grown. The first one is from a solution of the solute and the second from a melt of the substance. In addition there are several specialised techniques which are discussed towards the end of this chapter.

1 Growth from solution

There are four general methods by which crystals may be grown from solution. The first is by cooling the solution until a saturated solution is obtained (Fig. 1.6) and then forming a crystal around a seed crystal as in the above method. The second is to reduce the total amount of the solvent by evaporating the solvent and thus precipitating the solute, or making the solution more saturated. This technique is known as the evaporation technique but is rarely used except for when a selection of seed crystals is required.

Circulation in a temperature gradient. This method has recently been used a great deal for growing crystals of ethylene diamine tartrate and ammonium dihydrogen phosphate which are often given the symbols EDT and ADP respectively. These crystals are used as piezo-electric crystals. In Fig. 1.6 we have illustrated the apparatus used to prepare crystals by the temperature gradient method. In this method the poor crystals of ADP are placed in the bottom of a tube down which is placed a column of glass which has two vent holes at the top. The outside of the central tube contains a heating jacket and when the solvent is heated the solution is pumped up the inside tube to the top. Then a saturated solution is formed and crystals form on the surface of the seed crystals. The solvent, less some of its solute, passes to the poor crystals and the cycle is repeated. The temperature gradient is such that the solution is warmest at the bottom of the inside tube and least at the top of the outside tube.

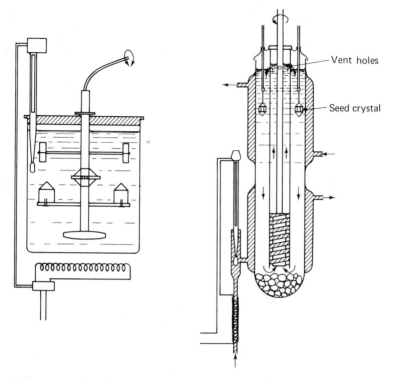

1. Temperature lowering method   2. Temperature gradient (circulatory)
   (a more advanced model of Fig. 1.4)   method

Fig. 1.6. Growth of crystals from an aqueous solution

Hydrothermal growth. Ruby, zinc(II) oxide, yttrium(III) iron(III) garnet and
aluminium(III) phosphate crystals may be prepared by this method. The
principle is that these substances dissolve in water at temperatures above
600 K but they do not dissolve at ordinary temperatures. We have seen that
solubility increases with temperature but the process is rendered very difficult
because very high pressures of more than several hundred bars (atmospheres)
are required to obtain water of this temperature 673 K (400 C). The boiling
point of water increases as the pressure is raised and conversely lowers as the
pressure is dropped. It is therefore necessary to use a very thick-walled vessel
for such experiments. In Fig. 1.7 a typical temperature gradient hydrothermal
apparatus is illustrated.

The nutrient material is placed in the autoclave together with the required volume of water. The seed crystals are then added and the top is
bolted down very carefully. The whole apparatus is then stood in thermal
insulation on a hotplate, so that the bottom of the autoclave is hotter than
the top. The nutrient material dissolves in the water and the solution rises
to the top and cools until it meets the seed crystal, on which the new crystal

12

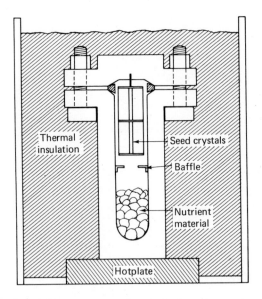

Fig. 1.7. Temperature gradient hydrothermal apparatus

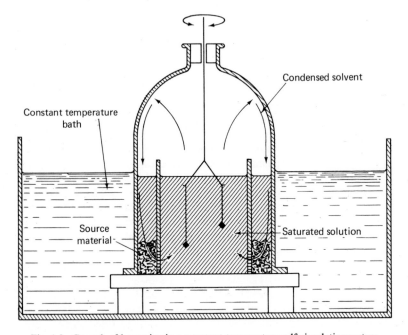

Fig. 1.8. Growth of hexamine in a constant temperature self-circulating system

is deposited. The limitations of the hydrothermal method are that the cost of equipment is very high and the experimenter cannot see the crystals forming. It is used when other methods to obtain crystals have failed. Corundum crystals may be grown in stainless steel autoclaves at temperatures up to 950 K, but there is a great danger of the lining of the autoclave walls dissolving in the solvent and precipitating in the forming crystal.

Self circulating system. Unfortunately ADP is naturally birefringent, which means that it has the property of being doubly refractive like calcite. Recently it has been shown that hexamine and KTN, which is a mixed potassium–tantalate niobate, are optically isotropic—that they have the same refractive index independent of the direction from which light is shone. In order to prepare hexamine a constant temperature self-circulating system as illustrated in Fig. 1.8 may be used. The apparatus is immersed in a constant temperature bath the temperature of which may be controlled in the same manner as the temperature of an aquarium. The apparatus consists of a sealed bell jar inside which the nutrient material is separated from the saturated solution by porous glass, through which the saturated solution may pass but not the crystals themselves. Inside the main compartment the crystals are slowly rotated and the saturated solution is always in equilibrium with the crystal. The system is self-circulating because the solvent distils from the central compartment to the compartment which contains the nutrient material.

In this way large single crystals of hexamine have been grown. KTN, as stated previously, is a mixed potassium tantalate niobate and is the most sensitive electro-optic material known. This means that its refracting properties change on application of an electric field, and such crystals are therefore able to modulate (or alter) the phase of a light beam.

2 Pure melt growth

Kyropoulos technique. In Fig. 1.9 the apparatus is illustrated. The molten material is kept just above its melting point in a crucible. A seed crystal which is cooled by water is touching the top of the molten salt and the seed crystal grows down into the melt. The rate of growth is controlled by the temperature of the furnace, and the crystals are usually annealed before being removed from the furnace. In

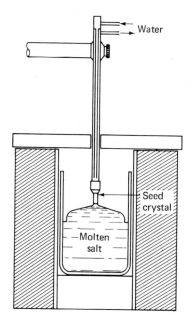

Fig. 1.9. Kyropoulos apparatus

14

the process of annealing the crystal is gradually cooled so that strains and stresses in the material are reduced. The main use for the method is in growing crystals of the alkali halides, for example sodium chloride and caesium iodide.

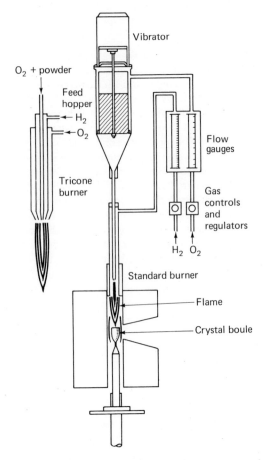

Fig. 1.10. Verneuil apparatus

Verneuil technique. Crystals of sapphires may be obtained from corundum which is a form of aluminium(III) oxide (Al_2O_3). In this technique hydrogen and oxygen are burnt in such a way as to melt the aluminium(III) oxide. The flame containing the oxide is directed onto a cold surface on which the molten oxide crystallises, the cold surface being gradually withdrawn yielding a single crystal of ruby if chromium(III) oxide is added. The aluminium (III) oxide is ground into a powder which is placed in a feed-hopper through which oxygen is passed carrying some of the powder into a regulated supply of hydrogen–oxygen mixture where the mixture is fused. The combustion of hydrogen in oxygen is a strongly exothermic process and this heat is sufficient to fuse the oxide which is directed onto a spinning boule of previously crystallised oxide acting as a cold surface on which the oxide solidifies. Note that sapphire, corundum and alumina are synonyms. The cold surface used in the melting techniques is not usually held firm but is either rotating, or spinning, or spinning and reciprocating which leads to boules (Fig. 1.11) of the required shape—as we shall see in Chapter 2 when the shape of crystals is discussed and in Chapter 6 when lasers are discussed.

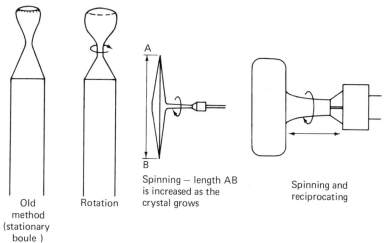

Old method (stationary boule)

Rotation

Spinning — length AB is increased as the crystal grows

Spinning and reciprocating

Fig. 1.11. Spinning boule technique

Plasma flame technique. A plasma is the region of very high energy produced by a high potential a.c. (alternating current) or d.c. (direct current) discharge. The powder is fed into the plasma, melts and condenses onto the cold surface. The advantage of the method is that very high temperatures can be reached and the more refractory oxides and carbides are prepared by this method.

Zone refining. Figure 1.13 shows the apparatus used for zone-refining or melting to obtain single crystals. In this method the polycrystalline material is maintained at a temperature just below its melting point, and then passed through an additional heating coil melting the material which, on moving through the flame, reforms as a single crystal. The formation of the single crystal depends on having a perfect seed crystal at the start of the process onto which the molten substance crystallises. The words 'single crystal' mean that there is but one crystal throughout the whole, whereas in a polycrystalline substance there are many crystals of the material present. The word polymorphic as opposed to polycrystalline means that a substance is capable of existing in two or more different forms. The technique is used for many crystals.

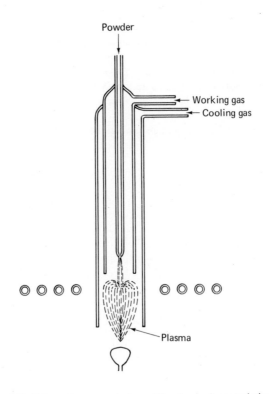

Fig. 1.12. Schematic representation of the plasma flame technique

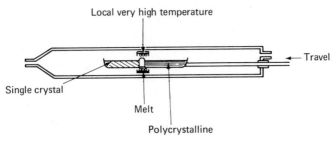

Fig. 1.13. Zone melting techniques

The Czochralski and 'pulling' techniques. The original method was designed to pull single crystals of metals from their melts. The method has now been developed more fully and millions of kilogrammes of silicon are produced each year for integrated circuits. The melt is maintained at just above the melting point (Fig. 1.14) and growth occurs as the heat is abstracted through the seed crystal. The seed is withdrawn at a rate which allows the interface between the solid and melt to remain in the same position. Nevertheless,

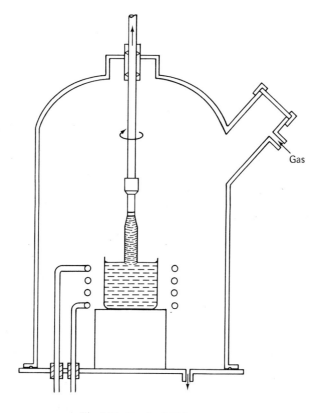

Fig. 1.14. Czochralski Apparatus

rapid growth rates of up to 50 millimetres per hour can be achieved provided that special precautions are taken to ensure the homogeneity of the crystal. The main precaution is to ensure that heat losses by radiation from the crystal surface are exactly balanced by local heating. The interface is consequently kept flat. In this way dislocation densities are minimised and strained regions avoided.

Germanium crystals of over five kilogrammes can be grown by this method provided high-purity carbon crucibles are used. Doped calcium tungstate crystals may be grown using platinum-alloy crucibles in an atmosphere of a noble gas.

3 Special techniques

The preparation of diamonds. In order to prepare diamonds pressures in the order of twenty to one hundred thousand bars (approximately atmospheres) and temperatures of the order of 3500 K are usually required. For a long time the efforts of chemists and engineers were frustrated by the inability to obtain these required conditions. The General Electric Company

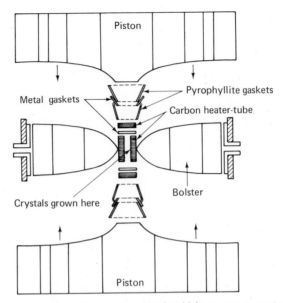

Fig. 1.15. Preparation of diamonds using ultra-high pressure apparatus

of America managed to combine these properties in an apparatus which is illustrated in Fig. 1.15. Essentially the apparatus consists of two opposing pistons which compress a very small region localised between two specially shaped bolsters. The apparatus uses pyrophyllite gaskets which are rather like talc at high temperatures. Only the region within the carbon heater tube is maintained at 3500 K and it is within this region that the crystals of diamonds are produced. The diamonds so produced are industrial ones and have approximately the same value as the natural variety, but they are a little harder wearing. The diamonds are used in cutting materials since they are composed of one of the hardest known substances.

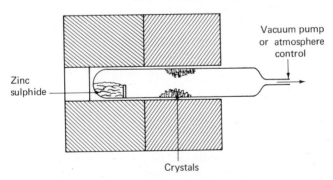

Fig. 1.16. Vapour deposition

Vapour growth. Crystals produced by vapour deposition are usually of very high purity, and if contamination occurs with a carrier gas the experiment may be carried out using a vacuum. In Fig. 1.16 the simple principles of the technique are illustrated. Crystals of zinc(II) and cadmium(II) sulphide may be grown by heating the substance so that it sublimes onto the cooler section. Sublimation is defined as the process in which a solid passes to a gas without passing through a liquid phase. Thus in this process the zinc(II) sulphide does not melt but passes immediately to the gaseous phase and then immediately back to the solid phase. Van Arkel prepared, for the very first time, a very pure sample of titanium after noting the fact that titanium-(IV) iodide (TiI_4) was unstable when heated and decomposed to titanium and iodine. The titanium(IV) iodide was placed in a vacuum in a vapour deposition apparatus and heated and the titanium was collected either on the surface of the vessel as with zinc sulphide or more usually on a hot filament. In this way the evaporation of tungsten filaments in tungsten-halogen lamps is limited. The technique of metal deposition is particularly important for the studies of the field-ion microscope and for the preparation of electron micrographs as we shall see in later chapters. The van Arkel technique is also used for other metals.

Experiment 1.5

A simple experiment that can be carried out at home or in a laboratory is to place some iodine or ammonium chloride in a test tube, heat it gently, and then sublimation can be seen.

How do crystals grow?

In order to answer this question we must be quite sure of the way in which atoms and molecules are distributed in the vapour, liquid and solid phases. In Fig. 1.17 the special points have been illustrated and to a large extent the order in the solid state is discussed in later chapters. The main feature,

Vapour	Random distribution of molecules. No region of order. Molecules often separated at large distances. Weak van der Waal forces of 20–40 kJ mol^{-1}. Often studied by electron diffraction but not X-ray diffraction.
Liquid	Region of order of molecules. Regions of disorder, molecules or ions separated 0·1 nm to 0·2 nm. At boiling point the intermolecular forces are overcome. Nuclear magnetic resonance studies are very useful but not X-ray diffraction.
Solid	Regions of order within crystal structures. Amorphous compounds, however, have little or no order. Ionic, covalent, coordinate and hydrogen bonds are important. X-ray diffraction studies often used to study regions of order.

Fig. 1.17. Distribution of atoms or molecules in solid, liquid and vapour phases

however, is that illustrated in Fig. 1.18 because in a crystal the atoms (or ions or molecules) are essentially very well ordered. In the liquid phase from which the solute molecules crystallise, the atoms or ions are less well

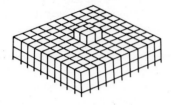 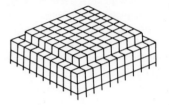

1. Atoms, ions or molecules build up in layers

2. To form new layers see Fig. 6.8

Fig. 1.18 One manner in which a crystal is formed

organised, and when a crystal grows from solution the atoms, ions or molecules become much more ordered. Very briefly, the property called entropy is a measure of the disorder in a system, and it is possible by finding the value of the entropy change to have a measure of the change of disorder going from a solid to a liquid, but the problem of entropy is discussed in more advanced university texts. There is a certain amount of evidence to show that metals and viruses grow in the layer form illustrated in Fig. 1.18.

The reader may experiment with organic chemicals like camphor which freeze like metals with small entropy changes, and these are being used to investigate a large number of solidification problems. These organic materials are capable of providing a visual method for trying to understand what happens when a metal solidifies, as the problem in the past has been the uncertainty of any theory which has been proposed. The materials which the reader might like to use are given in Fig. 1.19 together with their melting points and crystal structures. The meanings of the symbols FCC,

Material	Melting point *(K)	Crystal structure at m.p.
Adamantine	541	FCC
Borneol	483	FCC
t-butyl Bromide	248	FCC
t-butyl Chloride	245	FCC
Camphene	324	BCC
Camphor	448	FCC
Carbon tetrabromide	363	FCC
Cyclohexane	280	FCC
Cyclohexanol	296	FCC
Pentaerythritol	534	FCC
Succinonitrile	328	BCC

*K = C + 273.

Fig. 1.19. Organic crystals which behave like metals

etc., are discussed in more detail in later chapters. FCC means face centred cubic and BCC means body centred cubic. Briefly, the results have indicated that the layer method is one way in which crystals of metals and viruses grow. Normally a layer of atoms (or ions or molecules) is formed as the substance crystallises and this layer then acts as the foundation upon which the next layer may grow. The new layer grows in an ordered manner giving a complete layer, but faults arise in the crystal when a new layer is started before the old one is complete, such that there are gaps in the old layer.

Experiment 1.6

A simple mechanical analogy can be made if marbles are placed in a small box. Allow the marbles to completely fill the box, and it can be seen, as more are added, that the marbles fill up the holes in the layers below until a pattern of reasonable symmetry is formed. The reader may find it easier to pile the balls on one another in an open area.

Questions

Here are some simple questions which you might care to answer in two ways. First, write down from memory what you know, and second, write down what you know by reference to this book and any other books you may have in your possession.

1 Write down as many ways as you can of how crystals are used.
2 Write down as many ways as you can in which crystals are prepared.
3 How are metals used in colouring church windows?
4 What distinguishes a good diamond from a bad diamond?
5 How many examples of crystals can you find in the kitchen?
6 In what ways do diamond and graphite differ?
7 What is meant by the terms solubility, solubility curve and seed crystal?
8 What is meant by the term 'single crystal'? In what ways are single crystals important in the laboratory and in industry?
9 If crystals do not grow in layers can you work out any ways in which they might grow?
10 Why is it advantageous to grow crystals slowly?
11 What do you think is meant by the term 'twinning' of crystals?
12 What is meant by the term 'imperfections' in crystals?
13 What is meant by the term 'a supersaturated solution'? What is meant, consequently, by the term 'supercooling'?
14 There are two general methods of preparing crystals. Can you write down the names of these, and the names of the processes used under them?
15 The next chapter deals with the external structure of crystals. Can you, in preparation for it, answer the following questions?
    (a) What do you think are the different shapes of crystals?

22

(*b*) In what way might it be possible to classify crystals?

(*c*) Do you think the external structure of crystals is related to their internal structure?

References

BROWN, K. W., *et al*., 'Synthetic sapphires', *G.E.C. Journal*, **13**, 1944, 2.

BUCKLEY, H. E., *Crystal Growth*, Wiley, 1951.

GILMAN, J. J., *The Art and Science of Growing Crystals*, Wiley, 1963.

BUNN, C. W., *Crystals: Their Role in Nature and in Science*, Academic Paperback, 1964.

HUNT, J. D., *Acta Metallurgica*, **13**, 1965, 1212.

WHITE, E. A. D., 'The synthesis and uses of artificial gem stones', *Endeavour*, **82**, 1962, 73.

WOOSTER, N. and WOOSTER, W. A., *Mineralogical Magazine*, **29**, 1952, 858.

2 The external appearance of crystals

Are there any simple tests that can be applied to the study of the external appearance? There are many visual tests that have been applied, for example one can test how crystals cleave, and see how they appear under microscopes or electron microscopes. The angles between the crystal faces are often measured but one of the simplest qualitative tests is to test the hardness of the crystal.

MOH's hardness scale

Not all crystals are of the same hardness. For example, in the last chapter it was learned that diamonds are much harder than graphite, and similarly diamond is harder than apatite or talcum powder. An attempt was made to write down qualitatively the degree of hardness of the more common crystals. In Fig. 2.1 the scale is illustrated so that the softest crystal in the scale is talc which is used in talcum powder and the hardest is diamond. Each substance is placed in ascending order of hardness and the hardness is given a whole number or an integral value. The scale correctly shows that diamond is harder than topaz or apatite, but is wrong in giving the impression that topaz is eight-tenths softer than diamond or that apatite is half as hard as diamond. The hardness of a crystal, therefore, is not a good way of discovering which crystal one has. For example, if one had two crystals which looked the same it would not be a good way of distinguishing between them to measure the hardness. We shall see that a much better way to distinguish between two crystals is first to have a look at their external appearance and then their internal structure.

Substance (common name)	Hardness number	Chemical
Diamond	10	Carbon
Corundum	9	Aluminium(III) oxide Al_2O_3
Topaz	8	$Al_2SiO_4 \cdot (F \cdot OH)_2$
Quartz	7	Silicon(IV) oxide SiO_2
Felspar	6	Alumino silicate plus a base
Apatite	5	$Ca_5F(PO_4)_3$
Fluorite	4	Calcium(II) fluoride CaF_2
Calcite	3	Calcium(II) carbonate $CaCO_3$
Gypsum	2	$CaSO_4 \cdot 2H_2O$
Talc	1	$Mg_3Si_4O_{10}(OH)_2$

Fig. 2.1. Moh's hardness scale

Do all crystals have the same external structure?

Briefly, the answer to this question is 'No' and we can see why the answer is in the negative by having a look at some crystals. In Plates 2.1 to 2.5 several crystals are illustrated, and in the first there are some carefully grown crystals of bismuth. The crystals were prepared by heating bismuth (Plate 2.1) to just above its melting point and then cooling slowly. Close

Plate 2.1. Some crystals of bismuth

examination of the crystal shows that it is based upon a series of faces which join to form rhombohedra ($a = 0.4746$ nm, $\alpha = 57.2°$ (0.9983 rad) space group $R3m$—see later). If the crystal faces are tapped gently pieces of the crystal break or cleave along the planes in the crystal, the line along which the crystal is broken is called the cleavage plane. The second picture shows examples of crystals which were grown during research for maser and laser materials (Plate 2.2). These crystals were grown at the Mullard Research

Plate 2.2. Crystals used for lasers
Examples of crystals grown for research into laser materials

Plate 2.3. The surface of antimony metal (etched)

Plate 2.4. A life size sheet of aluminium which has been annealed and etched

26

Laboratories and crystal boules like these were used in the Goonhilly Downs Relay Station. They are ruby boules and were grown by melting aluminium oxide and rotating the cold surface onto which the stream of molten aluminium oxide was directed. The plates on p. 25 are of antimony (Plate 2.3) and aluminium metals (Plate 2.4). They were grown by heating to just above melting point and cooling very, very slowly. The cooled surface was then etched with a dilute acid to show, in the case of antimony, the dendritic or flowerlike patterns. A close examination of the dendrites shows that the lines join approximately at right angles and antimony has also a trigonal ($a = 0.407$ nm, $\alpha = 57.1°$ (0.9966 rad) space group $R3m$) symmetry like bismuth. The sheet of aluminium was annealed and etched with dilute acid and the light and dark patches show how the surface of the aluminium reflects light.

When an aeroplane with a propeller travels at high speed, the temperature of the metal approaches that of its melting point. When the aeroplane is not flying the metal cools slowly. The process is repeated for each flight and stop until eventually the metal forms a series of extended crystalline parts and cleavage can occur along the cleavage planes of the crystal. With continued heating and cooling the metal becomes annealed, a process in which the small regions of crystallinity join up to form larger regions of crystallinity. The crystals are able to break along the cleavage planes and hence the continued use of metal leads to metal fracture.

Plate 2.5 shows the urea formaldehyde insecticidal lacquer or DDT crystals grown from a 5 per cent solution in kerosene, in which multiple twinning can be seen. In the first chapter we recognised that the main

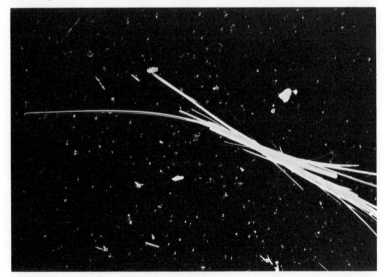

Plate 2.5. Urea formaldehyde insecticidal lacquers
DDT crystals form 5 per cent solution in kerosine

problem of growing crystals was in preparing single crystals. The problem
is often complicated by twinning. Twinning is not really the joining together
of crystals of the same substance. More correctly there are several forms of
twinning and the most common is the growth of two (or more) individual
crystals from one (or more) twin plane(s) in differing but symmetrically
related orientations. The most important property of twinning is that
twinned crystals often (but not always) show re-entrant angles. The crystals
shown have not the simple trigonal symmetry as of bismuth and
antimony and we are led to the important conclusion that crystals have
differing shapes. Each shape, as we shall learn is not only characteristic of
a particular substance, but often the internal structure is unique.

It is not easy to photograph crystals unless one has a single lens reflex
camera, but even without this it is possible to photograph some crystals if
you have them at home. The film ratings (Fig. 2.2) used for several of the
pictures was ASA 125 with a light value of five to seven with exposures from
one-sixtieth to half a second and stops of f/2.8 to f/8. When photographing
crystals each photograph is a new and separate problem to that particular
crystal. The presence of daylight helps to facilitate the lighting problems,
which greatly aid the appearance of the photograph.

Ilford FP3 (A.S.A. 125)   Fine grain

Substance	Light Value	Exposure	Aperture
Galena	6	$\frac{1}{2}$ sec	F.8
Calcite	7	$\frac{1}{60}$ sec	F.2.8
Glass	5	$\frac{1}{2}$ sec	F.8

Exposure on the specimen, not the background.
Glass is not a crystal but may be regarded as a super-cooled
liquid.

Fig. 2.2. Typical exposures for crystals

How do crystals appear under a microscope?

Optical microscope

Basically there are two types of microscope both of which are illustrated
in Fig. 2.3. The first type is the normal type of optical microscope which
can give a magnification up to fifteen hundred times. Using such a microscope it is possible to investigate the structure of some crystals. For example
in Plate 2.6 the single copper crystals or whiskers are compared with the
larger human hair. Whiskers were developed in an attempt to find materials
which were very strong along their length and occupy a very small volume.
The filaments or whiskers are often free from flaws or imperfections in

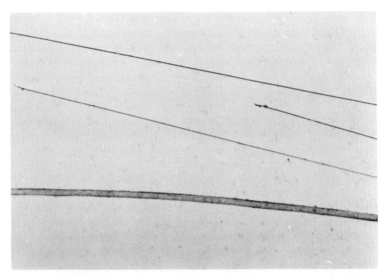

Plate 2.6. Single crystal of copper 'whiskers' and a human hair (the broader band)

Plate 2.7. The structure of a foam illustrated by a photograph.
Ideally the structure should be made up of regular pentagonal dodecahedra

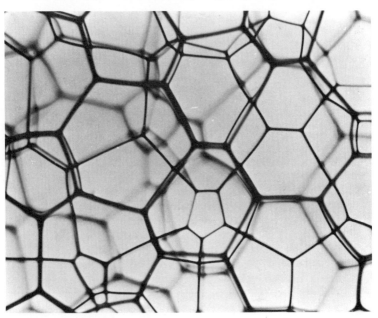

their structure and consequently are much stronger than ordinary metals. For example, one can calculate the expected tensile strength of a metal and very often the figure obtained theoretically is two thousand times stronger than that which is achieved in practice. Metals are much weaker than calculated as in many structures there are imperfections and dislocations, as we shall discuss later in this book. It is hoped that by introducing whiskers into the structure of a metal that the resultant composite with carbon will only be one fifth to one tenth as heavy as the ordinary metals. This reduction in weight will play a vital role in the development of aircraft and in rockets. The density of a whisker is exactly the same as the density of the parent substance from which it is made. But of course whiskers of carbon in the parent metal would reduce the average density giving a resultant loss in weight coupled with an increase in strength.

Plate 2.7 shows an unusual view of foam structure of the type that is met when a soap solution is agitated, for example when one is washing up. Essentially there is a framework or lattice which constitutes the structure of the foam. The framework is made up of the molecules of the foam and theoretically the structure should be made up of regular pentagonal dodecahedra but in practice there are only a few regular shapes. The whole structure of the foam is continuous in that all the molecules in the framework are in a continuous network. In a similar manner the structures of crystals consist of three dimensional arrangements of atoms (or ions or molecules) and the arrangement is characteristic of the substance.

The electron microscope

While the ordinary optical microscope is adequate for most studies on the form and structure of crystals, it is unable to show detail smaller than the wavelength of light. Objects such as viruses and molecules are smaller than this limit (about 1/10 000 mm) and can never be seen by normal microscopy. However, the development of electron microscopes now allows pictures to be taken showing details as small as a few Ångstrom units—a thousand times smaller than anything seen with microscopes using light to illuminate the object.

In the electron microscope a stream of electrons is emitted from the cathode, which is a hot vee-shaped tungsten filament, and accelerated by a high voltage applied to the anode A and passed through a special condenser lens onto the object. The electrons form an image at I which acts as an object for the new projector lens and an image is formed on the screen. The nature of the lenses does not concern us here but briefly they are magnetic fields designed to bend the electron beam, in the same way that a convex glass lens bends the light beam in the ordinary microscope. The screen may be replaced by a photographic film. The electrons can form a silver image on the film just as light does and, after the usual process of developing and fixing, a negative is obtained from which prints can be made. There is a high vacuum inside the electron microscope, as there is in a television tube, so that the electron beam is not scattered by gas molecules. In the microscope the vacuum is constantly maintained by special diffusion pumps and since the specimens go into the vacuum they have to be dried. The electron microscope can be used to examine both biological and metallurgical specimens.

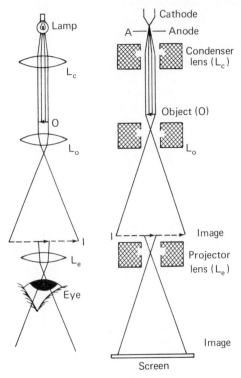

Fig. 2.3. The optical and electron microscope

The contrast in the image depends on the weight of the specimen. With metals, whose density may be around 20 g/cm³ the image of a small cluster of atoms only 0·5 nm may be dark enough to be seen. Biological material, on the other hand, only has the same density as water and is too transparent to be seen at very high magnifications. Special techniques have to be used to show small details. One is to combine a dense compound, like uranium-(III) acetate, with the specimen. Another is a method called shadow-casting in which the stream of a heavy metal such as gold or platinum is directed onto the specimen on the cellulose film such that the metal atoms arrive obliquely and form a thick layer on the 'windward side' and a clear space on the 'leeward' or shadow side. Electrons are unable to pass through the object where there is a thick layer of gold but are able to pass through the shadow where it contains no gold. The principle of the method is illustrated in Fig. 2.4.

Apart from the normal study of the surface of crystals electron micro-scopy is being used to study the arrangement of basic units in metal crystals and in the biological field in such objects as viruses and protein molecules. Insulin and haemoglobin, which are discussed in detail in Chapter 7, have

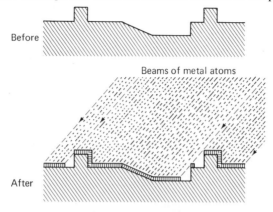

Before

Beams of metal atoms

After

Fig. 2.4. Shadow casting

been studied and such results as are available confirm X-ray diffraction results. In Plate 2.8 a sample of electro-deposited copper is seen to be rather like the Egyptian pyramids which is typical of metals with a cubic symmetry. So even polished metals like aluminium or tin are seen to be uneven when viewed through an electron microscope. The pyramid seen is about 30 nm high as each layer is one atom thick and the metallic diameter is about 0·3 nm. The bump is too small to be seen and too small to be felt. (1 nm is 10^{-9} metres so that 30 nm is 3×10^{-8} m which is about 10 000 times smaller than the smallest bump we can feel.) In order to study metal surfaces a solution of cellulose acetate is allowed to flow over the surface and it is possible to peel off a thin film which is an exact replica of the original relief of the surface. The replica may be shadow-casted if necessary and a typical electron micrograph is illustrated in Plate 2.9. Cutting tools are of necessity very hard and it is necessary to know the structure of the tip of the tool. A typical tip consists of the carbides of tantalum, titanium and tungsten and in the sample shown the crystals are not pure single crystals of the type previously discussed, but are now mixtures of the carbides in one another, or, more correctly they are said to be solid solutions of the carbides in cobalt as the solid solvent. In Plate 2.10 there is an enlarged print of an electron micrograph of a sample of china clay (kaolin) from Cornwall. Kaolin has been called kaolinite, nacrite, dickite and china clay and is a hydrous aluminium(III) silicate having the chemical formula $Al_3Si_4O_{10}(OH)_8$. The crystals belong to the triclinic system and the common form is as pseudo-hexagonal plates which being very soft crumble to a powder when pressed between the fingers. Under the ordinary microscope china clay seems like an amorphous powder but the

Plate 2.8. Sample of electrodeposited copper
Note: pyramidal growth. Magnification × 6000

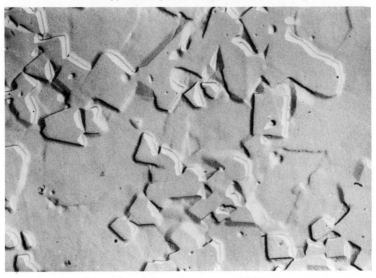

Plate 2.9. Electron micrograph of a cemented carbide material
showing mixed crystals of tantalum, titanium and tungsten carbide together with
fine grain tungsten carbide all in a cobalt binder. The polished surface was etched by
ion bombardment. This material is of the general type used for tips of cutting tools.
Magnification × 12 000 app.

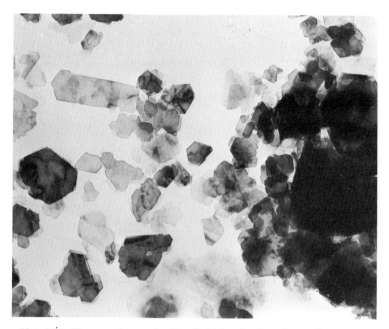

Plate 2.10. Electron micrograph of kaolin (china clay from Cornwall) × 21 000

electron microscope reveals the essential crystallinity of the chemical at a magnification of 21 000. The average size of the crystallites is 1μm (10^{-6} metres) in major diameter, but they are extremely thin. Most of the particles are seen to be aggregates in which several or many of the thin sheets are stacked, face to face, like a disorderly heap of playing cards. The aggregates can be largely separated by the action of dispersing agents for example 'Calgon' which is a polyphosphate used in an alkaline solution.

In plates 2.11 and 2.12 there are two different electron micrographs of the tobacco mosaic virus. In the first example there is part of a thin section of isolated tomato-fruit locule-tissue protoplast which had been obtained from a plant infected with TMV. The hollow centre of the virus is visible in this preparation and its tubelike structure can be determined. The material was stained for one hour with uranyl acetate, during dehydration, and the contrast of the virus is probably produced by the staining of the RNA core of the virus only. The material was embedded in butyl methacrylate/ styrene mixture (7:3). In the second there is a region of a crystalline inclusion of tobacco-mosaic virus from a freeze-etched, platinum-carbon replica, of a tobacco leaf specimen. There was no prior fixation with glutaraldehyde (as was the case with the thin-sectioned material) or treatment with an antifreeze which is usually a prerequisite for successful freeze etching. The crystalline structure can be seen and fracture has occurred both over the surface and within particles during the freeze-etching process; the magnification is × 110 000.

34

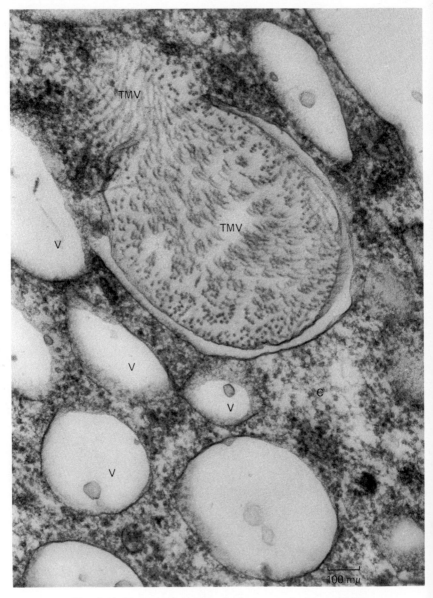

Plate 2.11. Electron micrograph of tobacco mosaic virus × 85 000
TMV—tobacco virus inclusion
V—cytoplasmic vesicle
C—cytoplasm

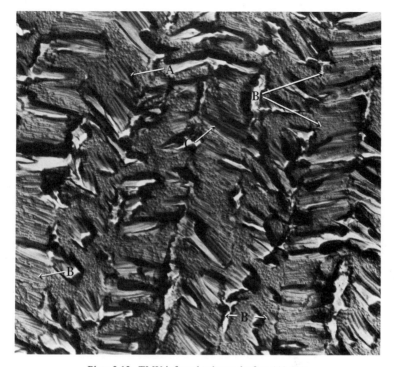

Plate 2.12. TMV infected tobacco leaf × 110 000
A—region where tubular structure of the virus is visible
B—regions where virus particles have fractured to reveal internal structures
C—surface view of whole virus rod

External form and the work of early crystallographers

Much of the work of early crystallographers was investigating the external appearance of crystals, together with some problems which concerned growing of different crystals. When sodium chloride is dissolved in a neutral solution the crystals obtained are cubic, but if sodium chloride is crystallised from a solution which contains a few crystals of urea the crystals are octahedral. The habit of urea is modified if grown from an aqueous solution containing some sodium chloride. Ammonium alum, $(NH_4)_2SO_4 \cdot Al_2(SO_4)_3 \cdot 24H_2O$, normally crystallises in an octahedral manner but those which are grown from an alkaline solution of sodium carbonate or calcium carbonate are predominantly cubic. N. Steno, who in 1666 studied quartz crystals, found that one can see that although the shape of the crystal does vary the basic difference between the crystals is that some faces are larger than others but the angles between the faces are constant. Steno then

36

formed his Law of Constant Interfacial Angles (Plate 2.13) which is formulated as 'The external shape or habit of a crystal of a given compound depends upon the relative development of the faces but the interfacial angles are constant.' This means that in the solutions of sodium chloride, alum or quartz for example, some faces are grown at the expense of others, and the angles between the faces are characteristic of the particular compound.

Plate 2.13. Alum crystals showing change of form and the law of constancy of crystal interfacial angle

A partial explanation in terms of the internal structure, which will be discussed in the next chapter, has been given by MacGillavry and co-workers. There is an epitaxial growth of a molecular structure in which the sodium chloride, urea and water are combined. The molecular structure forms and grows on certain faces which are then prohibited from growing because the parent substance sodium chloride or urea is unable to continue to build up the layers.

Plate 2.13 illustrates the changing form of some alum crystals. The cubic structure has small edges which can grow to give an octahedron.

The angles between faces are measured by instruments called goniometers, of which there are three types. The simplest is a contact goniometer which can be made at home by taking a protractor and a ruler such that the ruler passes through the centre of the protractor. Place the protractor along one face of the crystal and the rule along an adjacent face and measure the angle, as illustrated in Fig. 2.5. The next type is the reflecting goniometer and is more accurate than the contact goniometer. It consists essentially of a light source from which a beam of light is directed onto the surface of the crystal and the light ray is then reflected by the plane surface, and since the angle of incidence is equal to the angle of reflection the angle of incidence can be calculated. The crystal is then rotated through a known angle to give the new position in which the incident ray and the reflected ray are in the same positions as in the first case. The angle through which the crystal is rotated is the angle between the normals. This experiment works particularly well for the cubic crystals with good reflecting faces. The X-ray goniometer is used by crystallographers to measure the angle between the crystal planes and works on the same principle as the reflecting goniometer except that in this case the incident light is replaced by a beam of X-rays which are reflected and then detected by a method which will be discussed in Chapter 3. Apart from needing a high voltage to prepare them, X-rays are dangerous and must only be handled in a laboratory, using the right apparatus.

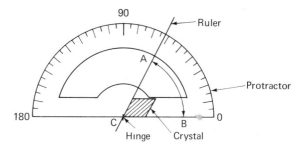

Fig. 2.5. Simple contact goniometer

Crystal symmetry

It was apparent to the early crystallographers that many crystals possessed varying amounts of symmetry and there are different types of symmetry, which concern the external shape of a crystal. These are now defined.

Plane of symmetry

A body has a plane of symmetry when it can be divided by an imaginary plane into two parts such that one part is the exact mirror image of the other. In Fig. 2.6 the planes of symmetry of cubic crystals are illustrated. If you are in any doubt about these planes of symmetry it would help either to draw them out yourselves or find or construct a cube and draw in chalk the planes of symmetry on the cube.

Rotation axis of symmetry

A rotation axis of symmetry is defined as the axis such that after rotation of $2\pi/n$ radians then the crystal, molecule or unit cell is exactly identical (where n is a small integer.) The axis is then said to be an $n$fold axis of symmetry. The axes of symmetry in a cube are illustrated in Fig. 2.7 and the

38

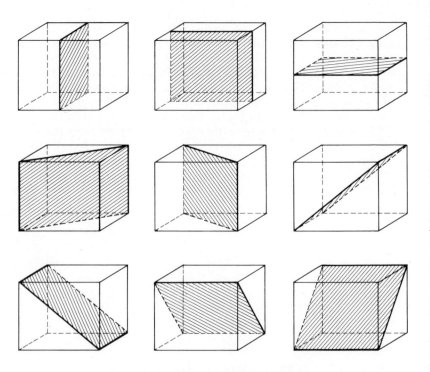

Fig. 2.6. The planes of symmetry of the cube
The planes may be regarded as surfaces of mirrors through which the two halves of the
cube are reflected. A point on one side exactly meets its equivalent when it is reflected

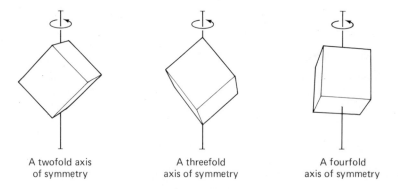

A twofold axis
of symmetry

A threefold
axis of symmetry

A fourfold
axis of symmetry

Fig. 2.7. Some axes of symmetry of a cube
A rotation axis of symmetry is defined as the axis such that after rotation through
360/n then the crystal, molecule or unit cell is exactly identical. The axis is then said
to be an n fold axis of symmetry

meaning of unit cell is discussed in Chapter 3, as the cell concerns the internal structure of the crystal. A twofold axis of symmetry is sometimes referred to as a diad axis, a threefold as a triad axis, a fourfold as a tetrad axis and a sixfold as a hexad axis of symmetry. Can you count how many planes of symmetry there are, how many twofold, threefold and fourfold, axes of symmetry there are in a cube?

Centre of symmetry

A centre of symmetry is defined as a point such that any line drawn through it will intercept the surface of the crystal at equal distances on either side.

(If all the numbers of elements of symmetry are added up then it will be found that there are twenty-three such elements of symmetry for a cube.)

The twenty-three elements in a cube are as follows:
3 tetrads along [100] [010] [001] directions
4 triads along [111] [111] [111] [111] directions
6 diads along [110] [110] [101] [101] [011] [011] directions
3 planes perpendicular to (100) (010) (001) planes
6 planes perpendicular to (110) (110) (101) (101) (011) (011)
1 centre of symmetry at $\frac{1}{2}\frac{1}{2}\frac{1}{2}$

—
23
—

Mathematicians and crystallographers realised that there are seven fundamental types of crystal or shapes and these shapes are listed in Fig. 2.8. The reader is encouraged to verify, by drawing, the Descartes–Euler theorem which states that

The number of faces + number of vertices = The number of edges + 2

This does apply for the dodecahedron eg $12 + 20 = 30(+2)$, but this is rather complex.

Type of solid	Vertices	Faces	Edges	Faces meeting at Vertex
Tetrahedron	4	4	6	3 Triangles
Octahedron	6	8	12	4 Triangles
Icosahedron	12	20	30	5 Triangles
Plane	∞	∞	∞	6 Triangles
Cube	8	6	12	3 Squares
Plane	∞	∞	∞	4 Planes
Dodecahedron	20	12	30	3 Pentagons

The Descartes-Euler Theorem states that faces + vertices = edges + 2.

Fig. 2.8. Solids and the number of external forces

Crystallographic axes and axial ratios

So far we have not really been at all precise; we have looked at various shapes and the different type of crystals that occur. A crystallographer is, however, a person who must be as precise as possible and it is therefore necessary to look into further detail. Crystals, in which the principal faces meet at right angles may be referred to orthogonal cartesian coordinates.

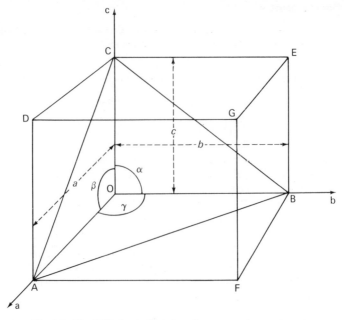

Fig. 2.9. The Millerian indices from intercepts on cartesian axes

In certain cases these coordinates may not be at right angles but wherever possible they are selected to be so. A standard or unit plane is selected, sometimes arbitrarily, but experience over the years has shown that the plane must intercept all three axes. In Fig. 2.9 the intercepts OA, OB and OC are given the symbols a, b and c. The unit plane is selected to have the smallest values possible of a, b and c. The absolute lengths of a, b and c are not critical but it is the ratio or the axial ratio which is of critical importance and all parallel planes have the same axial ratios. Normally it is convenient to write b as unity, thus for potassium sulphate

$$a:b:c = 0{\cdot}5727:1:0{\cdot}7418$$

The angles between the axes are fixed by convention and the angles BOC, AOC, BOA are represented by alpha α, beta β and gamma γ respectively. The elements of a crystal are the values of a, b and c and alpha (α), beta (β) and gamma (γ) which define the shape of the crystal. Figure 2.10 illustrates the crystal structure of anthracene and naphthalene in which the values of a, b and c are given. When an angle is a right angle it is not normally quoted but in this case the angle AOC (β) is not a right angle (but $\alpha = \gamma = 90°$). Can you see the relationship between the position of OB and the angle beta? —the axis b does not contain β. Crystals which have the same basic structure (same ratio $a:b:c$ and $\alpha:\beta:\gamma$) are said to be isomorphous and this we shall discuss in more detail shortly.

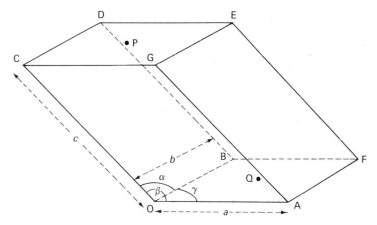

Fig. 2.10(a). The crystal structures of anthracene and naphthalene

	a	b	c	$\angle$ AOB $= \beta$
Naphthalene Å	8·34	6·05	8·69	122° 49′
Anthracene Å	8·58	6·02	11·18	125° 0′

$(\alpha = \gamma = 90°)$

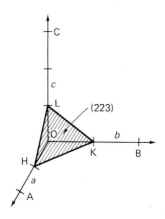

Fig. 2.10(b). Millerian indices of a plane

The Law of Rational Indices

The intercepts on a, b and c have specific values but it is not normal, to express a, b and c in this form and the usual way was devised in 1839 by Miller who used the relationship which connects the measurable angles to the planes or elements of symmetry. Sets of uniformly spaced planes can be drawn in an infinite number of ways but any given set of planes is defined by its spacings between the planes and its orientation to the three major axes. A set of planes consists of a series of uniformly spaced parallel planes. A set must, because the planes are uniformly spaced, divide the axes into a number of equal parts. Let the set dividing OA into h parts, OB into k parts and OC into l parts be called the (hkl) plane. Let this plane make an intercept H on the a axis, K on the b axis and L on the c axis so that a is divided into h parts each of length OH, b is divided into k equal parts each of length OK and c is divided into l equal parts of length OL. It is expressed mathematically as

$$\text{OH}:\text{OK}:\text{OL} = \frac{a}{h}:\frac{b}{k}:\frac{c}{l}$$

That is one refers to the Millerian indices h, k and l rather than a, b and c. The Millerian indices of a face are inversely proportional to the intercepts of that face on the chosen axes, thus if a is cut into two equal parts length OH, etc. (Fig. 2.10).

$$\text{OH}:\text{OK}:\text{OL} = \frac{a}{2}:\frac{b}{2}:\frac{c}{3}$$

and LMN is referred to as a (223) face and the indices are written in round brackets. When the values of h, k and l are unity the face is referred to as a (111) face. In future we shall be talking about orientation of crystal faces

---

The distance is written d_{hkl} where h, k, l are the Millerian indices. The expression is very complicated when the system is not orthogonal ($\alpha \neq \beta \neq \gamma \neq 90°$).

When $\alpha = \beta = \gamma = 90°$

$$\frac{1}{d^2_{hkl}} = \frac{h^2}{a^2} + \frac{k^2}{b^2} + \frac{l^2}{c^2}$$

when $a = b = c$ (cubic)

$$d_{hkl} = \frac{a}{\sqrt{(h^2 + k^2 + l^2)}}$$

Note: Planes (hkl) are written with brackets. hkl without the brackets means a reflection from the parallel set of planes (hkl).

Fig. 2.11 The perpendicular distances between planes (hkl)

to X-rays and the crystallographer chooses if possible the face with the lowest Millerian indices on which to direct the X-rays. If there is a negative intercept, for example $(1 -1 1)$ the intercept is usually written $(1\bar{1}1)$ with a bar over the symbol and it is quite possible to have such a face. For a perfect cube the crystallographic axes are chosen parallel to the three main edges and each face cuts only one axis, and consequently the symbols are (100), (010) (001), $(\bar{1}00)$, $(0\bar{1}0)$ and $(00\bar{1})$ for the different faces. This is a more convenient expression for these faces than writing down the ratios $a:b:c$ every time. Figure 2.11 indicates a method of calculating the perpendicular distance between (hkl) planes. In Fig. 2.12 some Millerian indices and the planes they represent are given. Can you see the relationship between the planes and the Millerian indices?

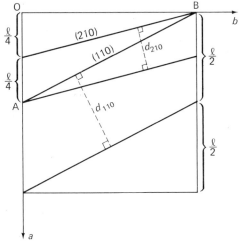

Fig. 2.12. Some Millerian indices and the planes they represent

Crystallographic systems

In the early nineteenth century it was recognised that there were different shapes of crystals, but nobody had really related these to the fundamental geometric systems. In 1830, however, F. C. Hessel used the Law of Rational Indices in a brilliant manner and showed that there were thirty-two different classes of crystal symmetry or point groups.

Point groups

Point groups are defined as the thirty-two ways in which the crystallographic elements of symmetry, that is centres, axes and planes may be distributed about one single point in space. Luckily these thirty-two classes fall into the seven fundamental groups, the names of which are given on the

Fig. 2.13 Crystallographic systems

System	Crystallographic elements	Essential symmetry	Number of point groups	Examples
Cubic, or regular	Three axes at right angles: all equal $\alpha = \beta = \gamma = 90°$ $a:b:c = 1:1:1$	4 triad axes; 3 diad, or 3 tetrad axes	5	CaF_2, ZnS, FeS_2, Cu_2O, $NaClO_4$ $NaCl$, Diamond, Pb, Hg, Ag, Au
Tetragonal	Three axes at right angles: two equal $\alpha = \beta = \gamma = 90°$ $a:b:c = 1:1:\frac{c}{b}$	1 tetrad axis	7	SnO_2, Sn, TiO_2, KH_2PO_4, $PbWO_4$
Ortho-rhombic or rhombic	Three planes at right angles: unequal $\alpha = \beta = \gamma = 90°$ $a:b:c = a/b:1:c/b$	3 diad axes	3	$PbCO_3$, $BaSO_4$, K_2SO_4, Mg_2SiO_4, α-S, KNO_3
Mono-clinic	Three axes, one pair not at right angles: unequal $\alpha = \gamma = 90°$ $\beta \neq 90°$ $a:b:c = a/b:1:c/b$	1 diad axis	3	$CaSO_4.2H_2O$, β-S, $K_2Mg(SO_4)_2.6H_2O$
Triclinic or anorthic	Three axes not at right angles: unequal $\alpha, \beta, \gamma \neq 90°$ $a:b:c = a/b:1:c/b$	No planes or axes	2	$CuSO_4.5H_2O$, $K_2Cr_2O_7$
Hexagonal	Three axes coplanar at 120°: equal. Fourth axis at right angles to other three $a_1:a_2:a_3:c = 1:1:1:\frac{c}{a}$	1 hexad axis	7	HgS, Ice, Graphite, Mg, Zn, Cd
Trigonal (Rhombo-hedral)	Three axes equally inclined, not at right angles: all equal $\alpha = \beta = \gamma \neq 90°$ $< 120°$ $a:b:c = 1:1:1$	1 triad axis	5	Calcite, Magnesite, $NaNO_3$, Ice, Quartz, As, Sb, Bi

Note: It is not good practice to use y for the ratio corresponding to c/b if x is used for a/b.

left of Fig. 2.13. The seven fundamental crystal types are closely related to the internal structure of a crystal, which subject is discussed in Chapter 3.

The question asked at the beginning of this chapter can be fully answered now. No, crystals are not all of the same shape, there are seven fundamental shapes. In particular, metals are sometimes cubic (sodium), but zinc and cadmium for example, are hexagonal and bismuth is trigonal. Ice crystals, which most of us have seen formed on a cold glass surface, are basically hexagonal like graphite. Calcium(II) sulphate, which is responsible for the hardness of water, is monoclinic. In later chapters calcium(II) fluoride is taken as a typical cubic ionic structure and zinc(II) sulphide as a typical cubic covalent structure. Rutile or titanium(IV) dioxide is tetragonal and is often taken as the typical example of this type of structure. The well known copper(II) sulphate has a very complicated crystal structure, being triclinic, and calcite which we shall discuss shortly, is trigonal.

Isomorphism

Having discussed the general shape of crystals we must now turn again to the problem of growing crystals. In 1816 Gay Lussac observed that a crystal of potassium alum ($K_2SO_4.Al_2(SO_4)_3.24H_2O$) will continue to grow when placed in a saturated solution of ammonium alum[$(NH_4)_2-SO_4.Al_2(SO_4)_3.24H_2O$]. That is, ammonium alum has the ability to form new layers on the crystal of potassium alum. If a drop of sodium nitrate solution is allowed to evaporate on a fresh cleavage face of calcite, which is calcium carbonate, small crystals of sodium nitrate are oriented so as to be closely parallel to the edges of the calcite cleavage. Similarly, potassium perchlorate ($KClO_4$) and potassium permanganate ($KMnO_4$) form parallel growths on barytes, which is barium sulphate ($BaSO_4$). When potassium phosphate (K_3PO_4) and potassium arsenate (K_3AsO_4) are grown separately or together the crystals formed are according to Mitscherlich apparently the same. In order to describe this similarity in chemical crystallinity Mitscherlich used the term 'isomorphism', and compounds which exhibit isomorphism are said to be isomorphous.

The Law of Isomorphism

The Law of Isomorphism states that the crystalline form of a compound is independent of the chemical nature of the combined atoms and is determined only by the number of atoms and by their method of combination. In the past this rule has been used a great deal, for example, since green chromic oxide was isomorphous with iron(III) oxide (Fe_2O_3) and alumina (Al_2O_3) the formula of green chromium(III) oxide was taken as Cr_2O_3. Again silver(I) sulphide was isomorphous with copper(I) sulphide (Cu_2S) and therefore the atomic weight of silver was found to be 108, that is, half the previously accepted value when the formula of silver(I) sulphide was taken as AgS. Nowadays it is recognised that two compounds will be isomorphous when they have the same formula type but the respective structural units, that is atoms or ions which will be discussed in Chapter 3, need not necessarily be of the same size in the two substances, although the relative unit cells must be rather similar in shape and size. The chemical properties of the substances should be rather similar.

Experiments

A quick look back through the chapter will indicate that there are many experiments to be done depending on the ingenuity of the experimenter. Thus pupils in one primary school devised their own hardness scale using flints, bricks, soil, sand, cement, etc., by measuring the distance that a nail sank into the material. In order to standardise the force with which the nail hit the substance the nail was put through a piece of wood which rotated about the opposite end. Crystals of metals can be made by laboratory versions indicated in Chapter 1 and a study of their external structure could be made. Cleavage planes may be studied by giving the crystal a sharp tap with a chisel. What angles do the cleavage planes make with the initial planes? Objects may be studied under a microscope or photographed as described. Why not look at a foam under a microscope or why not photograph the foam? Can you verify the Law of Constancy of interfacial angles? Grow crystals of sodium chloride in neutral and in alkaline solutions and in solutions containing urea and measure the angles between the faces with your own goniometer. Test the isomorphism of the compounds previously mentioned for example prepare crystals of potassium alum and then prepare a saturated solution of ammonium alum. Place the original alum in the new saturated solution and if the crystals are isomorphous the new alum will grow over the original crystal.

Questions

1 Define the hardness scale.
2 Describe how an electron microscope functions. Explain why the resolving power of the electron microscope is greater than the resolving power of an ordinary microscope.
3 Define the Law of Constancy of Interfacial Angles. How does this law lead to the classification of crystals according to their external structure?
4 Write down, with examples, the seven crystallographic systems.
5 Write down the systems of which diamond, quartz, fluorite, calcite and gypsum are examples.
6 How is the Miller index defined, and to what extent do the indices aid the classification of crystals?
7 Write down the elements of symmetry of a line, triangle, square and a cube.

8 Explain the following: (a) axis of symmetry; (b) plane of symmetry; (c) whiskers; (d) centre of symmetry; (e) twinning; (f) a polished metal may be rough.

9 What do you understand by the expression that light is a wave motion? Are there any other forms of radiation that may be regarded as having wave properties?

10 Write down concisely the way in which you think that the external symmetry of a crystal is related to its internal symmetry.

11 What is meant by (a) layer growth; (b) cubic symmetry; (c) crystal habit; (d) axial ratios; (e) Law of Rational Indices; (f) face (223); (g) crystallographic systems; (h) point groups; (i) sodium is cubic, zinc and cadmium are hexagonal and bismuth is rhombohedral; (j) isomorphism; (k) Cr_2O_3 is isomorphous with one iron oxide; (l) silver(I) sulphide is isomorphous with one sulphide of copper?

3 The evaluation of internal structure by X-ray diffraction studies

In the last two chapters we have seen how early crystallographers made empirical suggestions about the external structure of crystals. It was not until the turn of the twentieth century that crystallographers had any evidence about the internal structure of crystals, but let us briefly survey the historical development of crystallography. In the seventeenth century, Hooke suggested that a crystal may be built up from balls, for example, rather like lead shot or a pyramidal pile of cannon balls, and the collection of cannon balls formed the straight edges of the pyramid. In the eighteenth century, L'Abbé Haüy suggested that crystals might be made up of constituent molecules, each having the same shape as the bulk crystal. It is now known that Haüy was incorrect in the suggestion that molecules had the same shape as the crystal, but if instead of the word molecule the words 'unit cell' are substituted then he was right. The next advance was made by Mitscherlich when he defined his Law of Isomorphism. Then Pasteur found the geometrical enantiomorphism of dextro- and laevo-rotatory tartaric acid. Enantiomorphs are related by the fact that their external crystal structures are mirror images of each other, and enantiomorphism may be demonstrated by using the analogy that one's left and right hands are mirror images of each other. (Hold your hands so that the palms are touching and imagine that the mirror is held between the hands.)

In order to explain the chemistry of carbon Le Bel and van't Hoff postulated that the carbon atom in many compounds was surrounded, tetrahedrally, by four groups. On the basis of the tetrahedral carbon atom they were able to explain a great deal of the chemistry of carbon. In the nineteenth century, Hiortdahl and P. V. Groth tried to find relationships between crystal structure and chemical composition by altering regularly one atom in a series of compounds of the formula MX_2, for example, $MgCl_2$ and $CaCl_2$. They found that barium(II) fluoride BaF_2, strontium(II) fluoride SrF_2, calcium(II) fluoride CaF_2 all had the same structure, whereas magnesium(II) fluoride MgF_2 had a different structure, resembling rather more that of titanium(IV) dioxide or rutile. Again they found that lithium(I) nitrate, sodium(I) nitrate, potassium(I) nitrate, magnesium(II) carbonate and calcium(II) carbonate all have the calcite type of structure but one form of potassium(I) nitrate, and one form of calcium(II) carbonate, strontium(II) carbonate and barium(II) carbonate have the aragonite appearance (see Chapter 5). They found many typical changes in the external appearance of these crystals and, where there is a systematic chemical substitution in a series of compounds with similar chemical constitution, then the accompanying changes in structure are said to be morphotropic changes which are discussed in Chapter 4 with relation to the radius ratio effect.

In order to solve the problems met by the early crystallographers regarding the internal structure of a crystal it is necessary to discuss briefly the nature of wave motions, interference and diffraction and to understand the apparatus used by early crystallographers and modern crystallographers. At the beginning of this century a great deal of work was being done on different types of radiation in the electromagnetic spectrum (Fig. 3.1) from the very high energy gamma rays and X-rays to the lower energy radio waves. A close inspection of the complete spectrum shows that the visible light is but one small part of the total range. To the high energy side of this

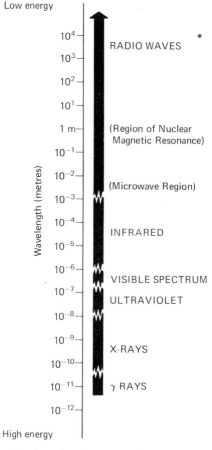

Fig. 3.1. The electromagnetic spectrum

light there is ultraviolet radiation, which causes sunburn, and further on there are the X-rays and then the γ-rays. X-rays are electromagnetic radiation of wavelength 0·1 nm (10^{-10} m) compared with an approximate value of (10^{-6} m) (1 μm) for visible radiation. Light waves may be diffracted by lines ruled on a polished surface provided that the lines are parallel and the distance between the lines is of the same order as the wavelength of light. Similarly we now know that X-rays may be diffracted by electrons on atoms provided that the internuclear distance is about the same as the wavelength of the X-rays. These rays are very dangerous as they are very penetrating

Fig. 3.2. Superimposing waves leads to interference patterns

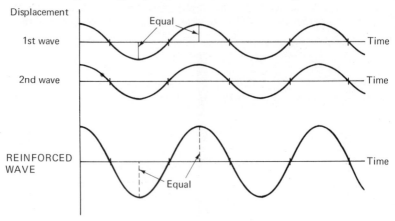

1. Complete reinforcement

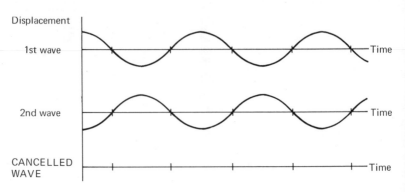

2. Complete cancellation

being able to pass right through human flesh so that anyone using X-rays must be very careful to use a lead screen to trap unwanted rays. The detection of X-rays is not difficult as they fog photographic plates but the rays are unaffected by electric or magnetic fields as they are not charged. Bragg used the fact that the high energy of X-rays causes them to ionise the gases through which they are passed. But let us return to the development of crystallography. Einstein showed that energy and frequency are related by the equation

$$E = hv$$

where E is the change in energy in ergs, h is the Planck Constant 6.6256×10^{-34} J sec, v is the frequency in sec^{-1}. The relationship that $c = v\lambda$ where c is the velocity of light, 3×10^8 m sec^{-1} (λ is the wavelength in metres) was well known. A particular type of radiation could be shown to have a wave motion if the radiation produced an interference or diffraction pattern in the way illustrated in Fig. 3.2. In the first case the first two waves are of the same phase and of the same wavelength and these completely reinforce one another when superimposed to give a wave of the same period but of twice the amplitude. In the second case there is complete cancellation because the waves now are of completely opposite phase and when these two waves are added together or superimposed they completely cancel each other.

The Laue method

Röntgen had produced X-rays but it was not known whether these had a wave motion or whether they were of a corpuscular nature as Newton proposed for light. In addition the crystallographers had the problem of not knowing the internal structure of crystals, and what was needed was someone to bring together the fields of crystallography and the study of X-ray radiation. Laue, who was a lecturer in physics at Munich where Röntgen produced the X-rays, had two research students, Friedrich and Knipping who considered that crystals might diffract X-rays in the same manner that a diffraction grating could diffract light rays.

The apparatus he used, or rather that which his students Friedrich and Knipping used, is illustrated in Fig. 3.3. In this apparatus the X-rays from an X-ray tube are collimated by lead screens making them pass through the crystal. The diffracted X-rays are collected on a photographic plate and are observed as a series of spots which correspond to the maximum reinforcement in the combination of the waves. So his students were able to

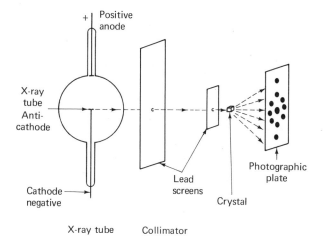

Fig. 3.3. Laue's experiment

show that X-rays were indeed wave motions, but the mathematics involved with his type of study of crystals was very complicated. In particular, the first results with copper(II) sulphate pentahydrate were very difficult to interpret because the X-rays did not all have the same wavelength and copper(II) sulphate has a low (triclinic) symmetry. When cubic compounds such as zinc(II) sulphide (zinc blende) and ammonium chloride were studied, the arrangements of the spots was much simpler, but, the method is very difficult when the arrangement of the atoms or molecules inside the crystal is required.

The Bragg method

The method was greatly simplified by Bragg when he decided to use crystals to reflect the monochromatic X-rays. Later in this book we will discuss the results of this method in detail, but if we regard crystals as consisting of layers of atoms or ions or molecules then X-rays may be reflected by these layers in the same way that light is reflected by the surface of the sea. It must be emphasised that crystals reflect X-rays in the Laue method. In the Laue method, however, the crystals remain stationary and each plane may reflect X-rays of different wavelengths. In the Bragg method the crystal rotates and reflects strong monochromatic radiation when the planes are at the correct angles to the beam. If the sea is rough then the light is diffuse, but if the sea is calm then a good reflection of the light occurs. Similarly, in a crystal when there is a regular arrangement of atoms then the X-rays will be reflected regularly, or, using the correct term, in a homogeneous manner. The X-rays are reflected by the electrons in the atoms and roughly the more electrons in an atom the more X-rays are reflected. That is, atoms such as iron which have a high atomic number readily reflect more X-rays than atoms of low atomic number such as hydrogen which has one electron. We shall see later that hydrogens are very difficult to detect in the presence of heavy atoms. Conversely (Chapter 7) it is often easier to spot a heavy metal atom such as iron in the presence of the much lighter atoms of hydrogen, carbon and nitrogen. In Fig. 3.4 the relationship developed by Bragg is illustrated. Two X-ray beams are incident on two layers of atoms or ions in a crystal at a glancing angle of θ. By simple geometry, AM is at right angles to MB and AN is at right angles to NB. Angles NAB and BAM are equal to θ because angle OAN $+ \theta = 90°$ and angle PAM $+ \theta = 90°$. The path difference between two X-ray beams is NB $+$ BM and

$$NB + BM = 2d \sin \theta$$

because NB $=$ BM $= d \sin \theta$. The rays will reinforce each other provided the path difference between the two rays is an integral number of wavelengths and the condition, therefore, for reinforcement is

$$n\lambda = 2d \sin \theta$$

n is an integer from one to infinity, and usually has the lower values. When $n = 1$ the difference in path length between the two rays is one wavelength and the interference or reinforcement is said to be of the first order. Similarly when $n = 2$ the difference in the path of two X-rays is two wavelengths and the reinforcement is said to be of the second order. λ is the wavelength of the X-rays and d is the distance between the atoms or ions in the

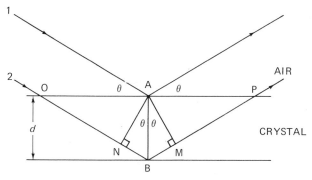

Fig. 3.4. The Bragg equation

layers. The Bragg equation is often used to find values of d, and values of θ can be found from the results of the experiment. The apparatus used by Bragg is illustrated in Fig. 3.5. X-rays from a source are collimated just as in the Laue method but instead of being passed through the crystal they are now reflected from the layers within the crystal itself. Then when the X-rays are reflected at certain values of θ, corresponding to the Bragg equation, the X-rays pass into the ionisation chamber in which a gas ionises. An

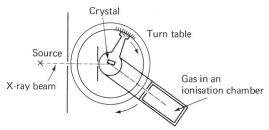

Fig. 3.5. The apparatus used (for example by Bragg) in some of the earliest X-ray diffraction experiments

ammeter connected to the ionisation chamber indicates when ionisation has occurred because no current flows until ionisation and by noting carefully the angular position of the ionisation chamber a value of θ may be found. Then, provided n and λ are known, a value for d may be calculated. But one question which we have not really studied is 'How do we know accurately the value of λ?' We shall return to this problem shortly.

The generation of X-rays

In a conventional X-ray tube X-rays are emitted as secondary radiation from an anode. Thus two types of radiation are produced, first a continuous spectrum which is analogous to white light, and secondly a characteristic line spectrum. (Fig. 3.6)

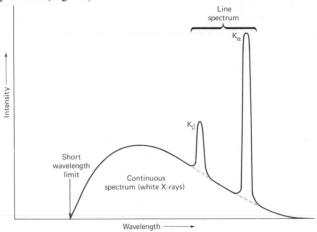

Fig. 3.6. Typical spectrum of X-rays

A The continuous spectrum

The energy associated with an electron of charge e which is accelerated through a potential of V volts is Ve. This is the quantum energy of the X-rays produced by a collision.

$$Ve = hv = \frac{hc}{\lambda}$$

where h is the Planck constant, c is the velocity of light and λ the wavelength. When the numerical values are substituted λ in nm is approximately equal to $124/V$. The value of λ (where V is in volts) is the shortest wavelength of X-rays produced. So roughly 124 000 volts are required to produce an X-ray of wavelength 1 pm. Just for the sake of completeness it is worth noting that λ was determined before e. The wavelength λ was first measured using a crystal of sodium chloride, by the method that we shall shortly discuss. W. L. Bragg correctly proved the structure of sodium chloride and potassium chloride and he proved that there were four sodium ions and four chloride ions per unit cell. All the evidence that he had was the Laue pattern; ρ = density; a = side of cube; M = molecular weight; N = Avogadro number. The side of the unit cell a is given by $\rho a^3 = 4M/N$ where $\rho M N$ are all known. Using this value of a, Bragg's ionisation chamber, monochromatic radiation, the formula $n\lambda = 2d \sin \theta$ gives a value for the wavelength because $1/d^2 = (h^2 + k^2 + l^2)a^{-2}$ and hkl were easily deduced from the spectra obtained. All the results fitted together like the pieces of a jigsaw puzzle.

B The characteristic line spectrum

The line spectrum is used as a source of X-rays of a known frequency because a single frequency (monochromatic) X-ray line may be selected using a filter. First, however, we must discuss the energy levels (Fig. 3.7) within an atom. These levels represent the energy of electrons from the lowest

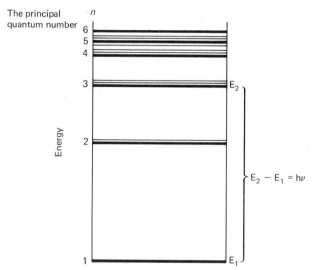

Fig. 3.7. The energy levels within an atom
$$E_2 - E_1 = h\nu$$

value when the principal quantum number n is one to higher values when n is six. The quantum numbers represent the order of energy of particular electrons. When considering very small standing wave motions like electrons it has been found that the energy cannot steadily increase. In fact, the energy is limited to a few values represented by the quantum numbers. The characteristic X-ray line spectrum is produced when the electron strikes the anode (which might be made of copper, silver or any other metal) such that an electron from an inner energy level may be removed. The vacant energy level is filled by another electron from a higher energy level and the energy change is accompanied by emission of an X-ray. By the Einstein relationship if the lower energy level is E_1 and the higher energy level is E_2 then

$$E_2 - E_1 = h\nu$$

where ν is the frequency of the X-ray. The jumps are lettered because when the lower principal quantum number is one the line is given the symbol K, when the lower principal quantum number is two the energy level is given the symbol L. The vacancy in the lower energy level may be filled by any one of a number of the higher energy electrons. One line which is often used is the K_α radiation from a copper target, the voltage required to produce K_α radiation from copper being $9kV$. The strong characteristic wavelengths emitted by copper atoms are $K_\alpha = 0\cdot154$ nm and $K_\beta = 0\cdot139$ nm. The K_β component is rather weaker than that of the K_α and the continuous background is even weaker. The copper K_β radiation may be absorbed by using a nickel filter because the K_β lines from copper are of an energy which can eject inner K electrons from nickel atoms. But the nickel in turn emits fluorescent X-rays of much longer wavelengths and which are scattered in all directions and not only onto the crystal. The effect is that the nickel filters out the K_β line.

Early determination of X-ray wavelengths

As a result of many X-ray determinations it was found that crystals consist of regularly repeating units, the smallest of which is defined as the unit cell, which will be considered later in the book. The characteristic spectrum obtained from the copper radiation has been of great value in the determination of the dimensions of the unit cell but first it is necessary to know the frequency or wavelength of the radiation. Historically this frequency was measured by reference to the known crystal structure of sodium chloride. For any unit cell

$$\text{Density} = \frac{\text{Weight of atoms in unit cell}}{\text{Volume of unit cell}}$$

$$= \frac{\text{Sum of the atomic weights of the atoms in a unit cell}}{\text{Avogadro's Number} \times \text{Volume of a unit cell}}$$

In the unit cell of sodium chloride there are eight sodium ions at eight corners and each corner ion is shared between eight cells, this gives an equivalent of one sodium ion per unit cell. In addition there are six sodium ions at the centres of six faces and each face is shared between two cells and this gives three additional sodium ions per unit cell. The total, therefore, is four sodium ions per unit cell and for electrical neutrality there must be four chloride ions in a unit cell. Using the above equation, to which we shall return later,

$$\text{Density} = 2\cdot16 \times 10^6 = \frac{4(23 + 35\cdot5)}{6\cdot2 \times 10^{23} \times a^3}$$

where a is a side of the unit cell, which is cubic, and the density of sodium chloride is $2\cdot16$ g/cm^3 and a is calculated to be $0\cdot59$ nm ($5\cdot9$Å). Substitution of this value into the Bragg equation for sodium chloride gives the value of the wavelength of copper K_α radiation, which we have seen to be $0\cdot154$ nm ($1\cdot54$Å).

X-ray diffraction apparatus and techniques

In Plate 3.1 a typical X-ray diffraction apparatus is illustrated. The X-rays come from the central tube and this is connected by four tubes or collimators to four different pieces of apparatus. The circular cylinder on the right-hand side is a typical powder camera, and on the lefthand side is a holder for a photographic plate by which the X-ray diffraction pattern from a single crystal may be measured. The large box on which the apparatus stands is a water-cooled apparatus by which X-rays are produced.

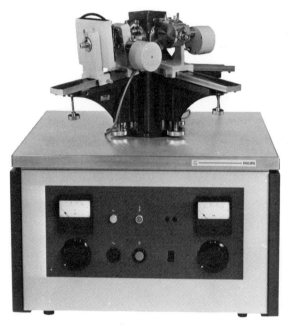

Plate 3.1 Photograph of equipment used in X-ray diffraction

The experimental techniques using X-rays are:

1 The Laue method

This method has been discussed previously and employs a narrow cylindrical beam of white X-rays (which contains X-rays of different frequencies) and the crystal and film are stationary. The photographs obtained show the crystal symmetry.

2 The rotation method

This is similar to the Bragg method. A narrow cylindrical monochromatic filtered beam of X-rays from a copper anticathode which has been bombarded with electrons is incident on a small single crystal rotating about a definite axis such as the face diagonal of NaCl perpendicular to the direction of the X-ray beam. The general type of apparatus used is illustrated in Fig. 3.8. The X-ray beam is collimated and either a cylindrical film or a plate film may be used. When a diffraction pattern is recorded on a plate or

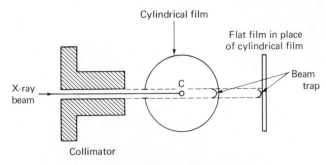

Fig. 3.8. Diagrammatic plan of the rotation camera

film the spots lie on hyperbolae, but on a cylindrical film with its axis coincident with the zone axis of the crystal the spots appear on straight lines. The diffraction spots lie on parallel lines which are said to be 'layer lines'. These photographs can be used to calculate the size of the unit cell but in the case of angular measurements because the spots tend to overlap and become indistinguishable a common method of refinement is the oscillation method which cuts down the number of spots.

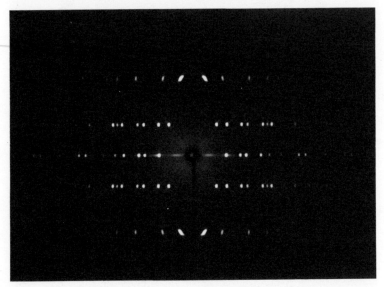

Plate 3.2 An X-ray diffraction pattern of urea viewed along the c axis

3 The oscillation method

This method is similar to the rotation method but now the crystal is rotated through five, ten or fifteen degrees, starting from a known orientation about a zone axis. The number of spots is limited and a whole range of zone axes is covered and several photographs (Plate 3.2) are taken instead of one. By this means better resolutions are obtained and such photographs may be used for complete structural determination. A typical X-ray diffraction pattern of urea, which has been oscillated about the c axis, is illustrated in Plate 3.2. The layer lines and the diffraction spots corresponding to maximum reinforcement are clearly seen. By measuring the intensity of each of these spots the complete structure of urea may be determined. The total results are then plotted in terms of a layer map of electron density which is computed from X-ray diffraction data. These contours of electron density are plotted on plates of glass which are separated at a distance, say of 10 mm, to represent a distance of 0·1 nm (1 Å) in the crystal. Rings are carefully drawn on the glass to represent regions of constant electron density. These are rather like the height contours of a mountain (see Plate 7.4). Where the electron density is greatest is considered to be an atom. A molecular model is then constructed of the crystal structure and then knowing the model of the molecule, the intensity of each spot is calculated from this model to see if they check accurately with the diffraction spots obtained. The model is refined until there is as good agreement as possible.

4 The Weissenberg method

This method is such that the apparatus is made complex and intricate but intensity measurements are simplified because the spots are separated. The film is screened so that only one layer line can be seen and the diffraction spots on that line are distributed about the whole film. The photographic film is made to move backwards and forwards parallel to the axis of rotation of the crystal, the rotation of the photographic film and of the crystal being synchronised so that one complete movement backwards and forwards corresponds to a complete 180° rotation. This kind of photograph provides all the necessary information required for the determination of the crystal structure and is often used to determine the structure of proteins and vitamins which will be discussed later in Chapter 7.

5 The powder method

This method is often used as a preliminary to, and in conjunction with, one or more of the other methods. It is particularly useful when single crystals are not available, and to obtain an identification of the particular substance under investigation. In Chapter 1 we saw the example where the girl was growing apatite in her hands. By comparing the known powder photograph of apatite with that of the apatite from her fingers, crystallographers were able to show that the two powder patterns were identical. A whole library of powder patterns has now been built up and it has been

60

possible to identify many compounds very quickly. The crystal specimen is powdered and gently eased into a capillary tube and in this tiny column of powder the crystals may be no more than 10^{-5} mm in size. The apparatus is illustrated in Fig. 3.9 and a monochromatic X-ray beam which is produced by the usual way is incident on the column of the sample and the pattern

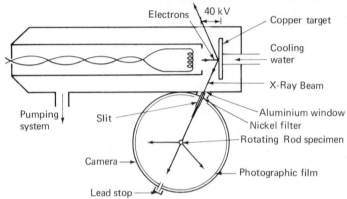

Fig. 3.9. The apparatus used for X-ray powder photography (not to scale)
The distance of the target to the filament is but a few millimetres and the slit must be very close to the window. The radius of the camera is approximately 10 centimetres. Without proper precautions this system would be very dangerous because of the escape of X-ray radiation. X-rays are in fact extremely dangerous and great care must be taken to cut out stray radiation. Often lead shields are placed around instruments

produced is recorded on a film which is wrapped round the inside of the camera. X-rays may pass straight through the crystal and are stopped by a lead stop. The rod specimen is rotated to give as many chances for the Bragg condition for reinforcement to apply. A typical powder pattern could be taken with an acceleration voltage of 40 kilovolts (kV) and a current of 20 milliamperes (mA) using a copper anode and a nickel filter, and is illustrated in Plate 3.3. There is a series of rings whose intensity and diameter are characteristic of the particular compound. When the crystals (10 μm diameter) are not finely divided, then the rings sometimes appear slightly spotty, but if the spots are diffuse, then the crystallites are very small indeed. It is possible to measure intensity either by the eye or using an instrument called a 'densitometer', which records the intensity of each line on a piece of paper, and the thickness of the line, but often cannot distinguish between small variations of intensity as the eye can. Where two compounds are isomorphous the powder patterns are similar, for example Fig. 3.10 illustrates two compounds which are isomorphous. The nature of the compounds does not matter but here one is a chloride of a palladium compound and the other is the bromide of a palladium compound. When the results are recorded on a densitometer and plotted against the diameter of the rings, the rings are seen to be approximately of the same intensity and of the same diameter. It is important when studying powder photographs

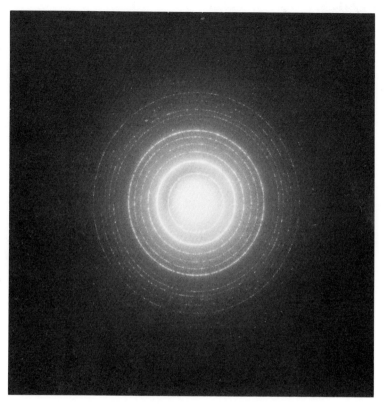

Plate 3.3. The powder pattern of thallium(I) chloride

to look not only at the intense rings of small diameter but also at the rings of low intensity and large diameter. The reason is that these rings will often tell you whether the compound is pure and also whether it really is the compound that you think it is. Where compounds are polymorphous, that is they exist in different crystalline forms, the powder patterns will be different; for example, sodium tripolyphosphate exists in at least three different forms and each form has a characteristic powder pattern. The powder patterns of metals are particularly interesting because when two pieces of metal are being investigated, if they have identical histories, very often they give identical powder patterns. But in cases where metals have been annealed or crystallised, or where a centre for fatiguing has developed so that the regions of crystallinity have joined together the powder rings are incomplete and indicate a relationship between crystallite orientation and the direction of mode of working.

Finally a word about the detection of X-rays. In the early days the positions and intensities of diffracted beams were found by an ionisation

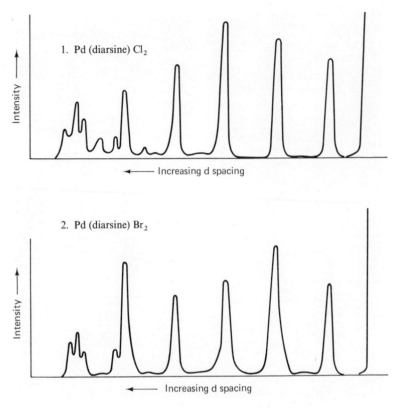

Fig. 3.10. Two isomorphous compounds, the powder patterns of which are represented by densitometer plots

method because X-rays have a high energy and can cause gases to ionise, but this was replaced by a photographic method. Recently, however, the Geiger counters are being widely used, especially in the more modern instruments.

Structure determination

From the preceding sections in this chapter it can be seen that X-ray structures are primarily concerned with determining the arrangement of atoms, ions, groups, or layers within a crystal. The chemical information required and obtained is very important. The kinds of atoms present and their groupings should preferably be known from previous chemical experience. For example, we shall see in Chapters 4 and 5 that a crystal may be either essentially ionic or essentially covalent and it is important to know which. Secondly, it is important to place the crystal in a particular classification, that is whether it is cubic, hexagonal, tetragonal, etc. Then by

Fig. 3.11. Dimensions and lattices

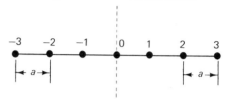

1. A one-dimensional lattice. Coordinates of points are $-3, -2, -1, 0, +1, +2, +3$

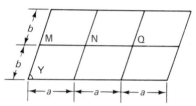

2. A two-dimensional lattice. Coordinates given by unit translations along a and b

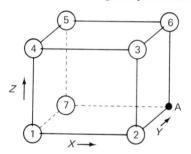

3. A three-dimensional lattice is made up of unit cells
Note the right hand rule is always used for the axes.
What are the coordinates of 1, 2, 3, 4, 5, 6, 7?
Which is the unit cell in the (a) one-dimensional lattice (b) two-dimensional lattice?
What examples do you know of (a) one-dimensional structure (b) two-dimensional structure (c) three-dimensional structure (d) four-dimensional structure?

examining the X-ray diffraction pattern it is important to be able to determine the arrangement of the atoms within the unit cell and certain absences in the diffraction pattern are indicative of certain types of symmetry. Where a crystal is very symmetrical there are few spots and the absence of many spots indicates a high degree of symmetry. A complex pattern is obtained when there are but few symmetry elements within a crystal or when there is a large unit cell. It is important, therefore, to understand the different types of symmetry that can occur within a crystal.

64

The crystal is built up from a framework of atoms or ions. In Fig. 3.11 if the lattice on which the structure is based were in one dimension then the crystal would be formed along one line and the coordinates of a point could be those illustrated. In two dimensions the coordinates are given by one or more unit translations along the axes a and b as illustrated for the classification of crystal faces in Chapter 2. In three dimensions the complete framework is made up of unit cells, which are defined as 'the smallest possible repeating unit based on a parallelepiped'. The coordinates of particular atoms or ions are rather similar to those we used for particular faces. For example, the point A in the three dimensional lattice has the coordinates 1, 1, 0. Can you write down the coordinates of 1 to 7?

Symmetry

In Fig. 3.12 axes of symmetry are related to the internal structure of a linear molecule AB. This only has a twofold axis of symmetry when A is the same as B but has an infinite-fold axis looking along the line A—B. In an equilateral triangle there is a threefold axis of symmetry about P, when A ≡ B ≡

Fig. 3.12. Axes of symmetry

1. A line has a twofold axis about the midpoint P (when A ≡ B)

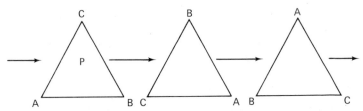

2. An equilateral triangle has a threefold axis of symmetry about P. How may point P be found by drawing?

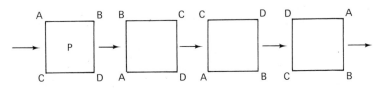

3. A square has a fourfold centre of symmetry about P which is the centre of the square

C. How can you find P by drawing? In a square there is a fourfold axis of symmetry about P which is the centre of the square when A ≡ B ≡ C ≡ D. Can you find any more elements of symmetry? In a planar AB₃ molecule or in the carbonate ion (Fig. 3.13), which is planar with a carbon at the

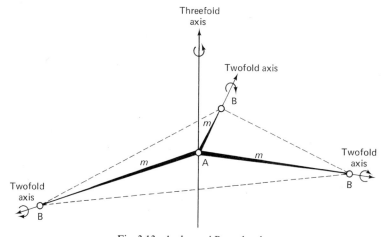

Fig. 3.13. A planar AB₃ molecule
The symmetry is one triad (threefold) axis and three planes of symmetry (*m* = mirror plane of symmetry) normal to the molecular plane, three diad (twofold) axes and one *m* in the plane

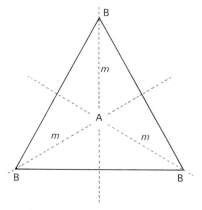

Fig. 3.14. The planes of symmetry in an AB₃ molecule. This is a plan view looking at the molecule. Each of the three planes of symmetry cuts through one B atom, the A atom and bisects a line B–B. There is normally no B–B bond

centre of the triangle of oxygens, there is one threefold axis and three two-fold axes. In addition there are three planes of symmetry which are parallel to the threefold axes which are illustrated in Fig. 3.14. Figure 3.15 is particularly important because it emphasises how a centre of symmetry is present within simple molecules. The structure of the phthalocyanines, before the structure of much simpler molecules, was elucidated because certain atoms, those of nickel, could be interchanged at the centre of symmetry, giving a series of isomorphous compounds. Where a heavy atom can

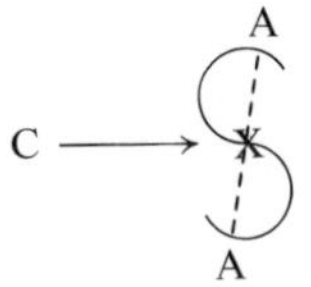

The figure S has a centre of symmetry at C. Any point A when inverted through C reaches its image.

xyz to $-x$, $-y$, $-z$ which are written $\bar{x}\bar{y}\bar{z}$

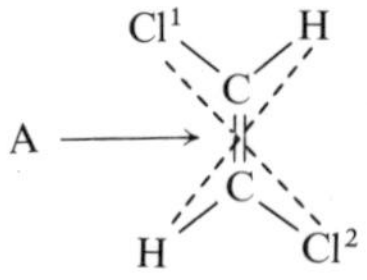

Trans 1, 2, dichloroethene has a centre of symmetry at A which is equidistant between the carbon atoms. If, say, a chlorine (Cl^1) atom was inverted through A it would reach its image (Cl^2).

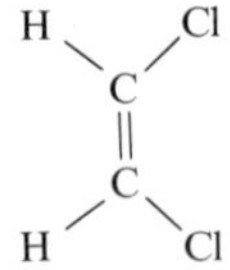

Cis, 1, 2, dichloroethene does NOT have a centre of symmetry
Can you see why?

xyz to $\bar{x}\bar{y}\bar{z}$ inversion is three dimensional

xyz to $\bar{x}\bar{y}z$ rotation is two dimensional

xyz to $\bar{x}yz$ reflection is one dimensional

The point through which inversion takes place is the inversion point or centre of symmetry. Rotation takes place about a line. Reflection takes place across a plane one-dimensionally.

Fig. 3.15. The centres of symmetry

be placed at the centre of symmetry then the X-ray diffraction study is greatly simplified.

The absences indicated that the heavy metal atom was at the centre of symmetry. One of the main difficulties with the interpretation of the diffraction patterns is in deciding the phase of the waves and this phase problem is greatly simplified when the heavy metal atom is at the centre of symmetry. We have already said that X-rays are reflected by the electrons and the heavy metal atoms really play a major role in the final intensity of the diffraction spots. There are two further types of symmetry, those of glide planes and screw axes which are rather more difficult to visualise, and these are illustrated in Figs. 3.16, 3.17, 3.18 and 3.19.

Although crystallographic nomenclature in the main is beyond the scope of this book it is important to realise that the presence of certain elements of symmetry means also that others must be present. Whereas a single mirror plane by itself does not imply a twofold axis, a mirror plane plus a centre of symmetry in the mirror plane creates, in addition, a twofold axis.

The nomenclature often used is the Hermann–Mauguin notation. For example, P means a primitive lattice, 2 means a twofold axis, 2_1 means a twofold screw axis, and a $P2_1$ means a primitive lattice with one twofold screw axis of symmetry, a $P2_1/m$ means a primitive lattice with one twofold screw axis and an additional mirror plane perpendicular to the principal axis, and often if there is a glide plane this is mentioned in the symbol at this stage. Then it is necessary to write down the axes of lower symmetry and mirror planes and glide planes which are perpendicular to these axes. Only

Fig. 3.16. Glide plane

These are reflections across a plane followed by translations or glides parallel to an axis in that plane such that a repetition of the two movements brings the whole system into coincidence.

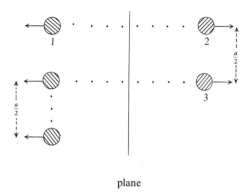

plane

Plane	Glide plane

Comparison between a plane and a glide plane

68

n is the order of the axis (n = 2, 3, 4, or 6)

p/n is the fraction of the unit lattice translation (the pitch of the screw) through which the screw translation takes place.

p/n is

A diad axis is 2_1

A triad axis is 3_1 or 3_2

A tetrad axis is 4_1 or 4_2 or 4_3

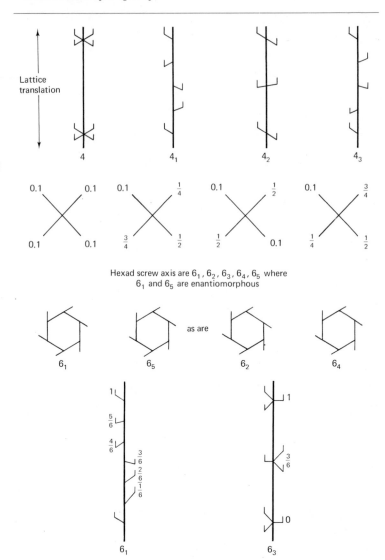

Hexad screw axis are 6_1, 6_2, 6_3, 6_4, 6_5 where 6_1 and 6_5 are enantiomorphous

Fig. 3.17. Screw axes, np

69

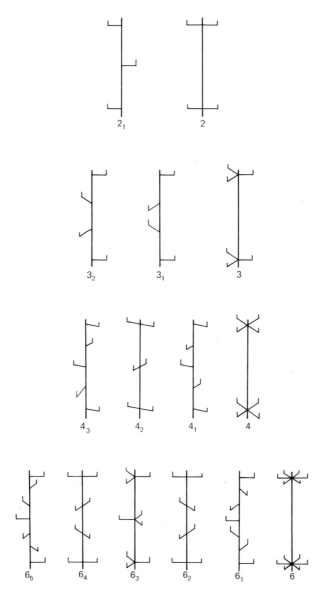

Fig. 3.18. Comparison of pure rotation axes and screw axes

70

Fig. 3.19. The fourfold inversion axis of symmetry of urea
A body having a fourfold inversion axis, which is called $\bar{4}$, is brought into coincidence
with itself by rotation through $2\pi/4$ radians about an axis plus inversion through a
point on that axis

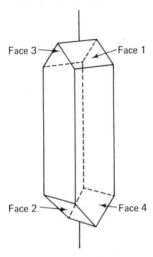

these axes of lower symmetry, mirror planes and glide planes are necessary
to define the point group or space group uniquely. P2/m2/m2/m is an orthorhombic group with twofold axes which are mutually perpendicular with
mirror planes at right angles to each of the three twofold axes. The abbreviated form Pmmm is often used because the twofold axes are implied by
the mmm, but note that 222 does not imply mmm and 222 is a different
point group.

The symmetry elements for all space groups are given in international
tables of X-ray crystallography.

---

1. P = primitive lattice.
2. $P2$ = primitive lattice with only a twofold axis.
3. $P2/m$ = 2 with an additional mirror plane perpendicular to the principal axis
   (glide planes are also mentioned now).
   Next following axes of lower symmetry and mirror planes or glide planes
   perpendicular to these axes.
4. $P2/m.2/m.2/m$ = orthorhombic group (with twofold axes which are mutually
   perpendicular), at right angles to each axis there is a plane of symmetry.

When certain elements of symmetry are present it is shown that others must be
present. A mirror-plane is equivalent to the twofold inversion axis ($\overline{\text{m}}$s $\bar{2}$) but this does
not mean that a mirror plane m involves a twofold rotation axis. Two-mirror planes
intersecting normally do mean that the line of intersection is a twofold axis ($mm2$). The
abbreviated form Pmmm is invariably used. The symmetry elements for all the space
groups are given in International Tables for X-ray Crystallography.

Fig. 3.20. Simple crystallographic nomenclature (Hermann-Mauguin notation)

The internal structure of crystals

The space lattice

The word lattice is often misused since lattices are often regarded as being synonymous with structures. Shortly we shall talk about the Bravais lattices which are the theoretical ways in which points could be distributed. One unit cell by itself cannot be a lattice which is considered to be infinite. The unit cell is the basic three-dimensional unit of pattern which when repeated by the lattice translations gives the whole crystal framework. In terms of the lattice itself the unit cell can be said to be the unit set of lattice translations, including both lengths and relative directions, completed to form a parallelepiped. One of the best approaches to lattice points is that they are reference points having identical surroundings. Imagine being at one reference point and then on being translated to another lattice point the surroundings or view would be the same.

The distances $a\ b\ c$ are called the primitive translations and represent the lengths of the chosen unit cell.

The Bravais lattices

Long before X-ray diffraction studies were made on crystals, A. Bravais in 1848 showed, using arguments of pure geometry, that there were only fourteen kinds of simple space lattices. These are illustrated in Fig. 3.21. The symmetry of these simple space lattices exactly corresponds to the seven crystallographic systems which were classified according to the external symmetry. The information which may be found in crystal structure tables consists of the angles ($\alpha\beta\gamma$), and the lengths (abc) of the unit cells which are always taken to be parallelepipeds. From these figures the volume of the unit cell may be calculated (V_{uc}). The numbers of atoms (Z) per unit cell are all quoted from which the molar volume (V_M) may also be calculated:

$$V_M = \frac{6.023 \times 10^{23}}{Z} V_{uc} \quad (\text{Å})^3 \text{ or nm}^3 \times 10^{-3}$$

$$= \frac{0.6023}{Z} V_{uc} \quad \text{cm}^3/\text{mole}$$

Since

$$\text{Mass} = \text{Volume} \times \text{Density}$$

$$\text{Density} = \frac{\text{Mass}}{\text{Volume}}$$

$$\rho = \frac{M}{V} = \frac{\text{Molecular Weight}}{\text{Molar Volume}}$$

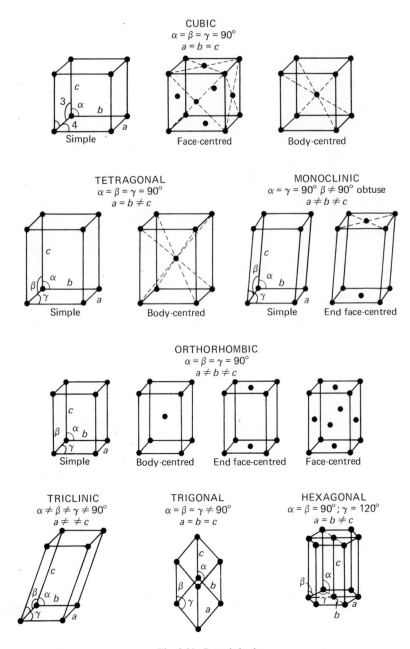

Fig. 3.21. Bravais lattices.
α is the angle between positive b and positive c. β is the angle between positive c and positive a. γ is the angle between positive a and positive b.

The 'mass' that we require is the molecular weight (M) and the volume is the molar volume (see chapter 6)

$$\rho = \frac{MZ \times 1.6604}{V_{uc}} \text{ g/cm}^3$$

It may be shown that the volumes of the unit cells of the seven crystals system are special cases of the formula

$$V_{uc} = abc \, (1 + 2 \cos \alpha \cos \beta \cos \gamma - \cos^2 \alpha - \cos^2 \beta - \cos^2 \gamma)^{\frac{1}{2}}$$

There are three types of cubic structures, two types of tetragonal structures, two of monoclinic, four of orthorhombic, and one each of triclinic, rhombohedral and hexagonal. These are really the simplest unit cells of the seven different crystallographic systems and one must be very careful in determining the number of atoms or ions per unit cell. At first sight it might be thought that there are eight ions per unit cell in the simple cubic system but in fact there is only one per unit cell because each of the eight atoms or ions are shared by eight other unit cells and so, on average, there is one atom or ion per unit cell. Similarly for the body-centred cubic structure there are two atoms per unit cell. We have already discussed the number of atoms per unit cell in the face-centred cubic structure when we talked about the density of sodium chloride in relation to the determination of the wavelength of X-rays and we shall return to the problem when we attempt to calculate the percentage free space on a metal.

The space groups

The space groups Fig. 3.22 are the infinite systems obtained by adding the lattice translations to the point group symmetries, in the maximum number

Crystal system	Space groups
Cubic	36
Tetragonal	68
Hexagonal	27
Rhombohedral	25
Trigonal (Orthorhombic)	59
Monoclinic	13
Triclinic	2
Total	230

Fig. 3.22. Distribution of the space groups between the seven fundamental systems

of ways. It is important to realise that the number of types of symmetry elements involved in the point groups is very small. The rotation axes 2, 3, 4, 5, and the inversion axes $\bar{3}$, $\bar{4}$, $\bar{6}$, the plane of symmetry and the centre of symmetry are simply combined in different ways to give the point groups.

To obtain the space groups one only has to add the translations. In all there is a total of 230 types of spaces groups which are distributed among the fourteen space lattices. For example, zinc blende and carbon have the same space lattice but because diamond is made up completely of carbon and zinc blende is made of zinc sulphide units then the symmetry of the two compounds is different. Zinc blende crystals have lower symmetry than diamond and are in a lower space group.

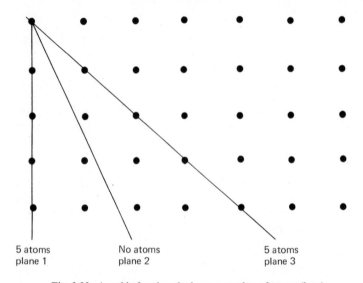

5 atoms   No atoms   5 atoms
plane 1   plane 2   plane 3

Fig. 3.23. A stable face has the largest number of atoms (ions)
Planes 1 and 3 could be crystal faces (1 being more probable than 3); plane 2 could not grow (although it could be cut on, say, a diamond) as a natural face

We are now in a position to be able to see how the external structure is related to the internal structure. Cleavage occurs along the plane which contains the maximum number of atoms or ions, and along a region where the crystalline forces are weakest. These forces are weakest when the reticular density (network density) is highest and the plane-spacing the greatest. The term reticular density, however, is very difficult to define and it is doubtful whether it has real meaning in any but the simplest crystals. A stable face (Fig. 3.23) has the largest number of ions. Planes 1 and 3 are stable, each containing five atoms or ions but plane 2 is unstable because it does not contain any atoms or ions. Naturally growth of crystals occurs at the outermost surface of the crystal and if the surface consisted of positive and negative charges these surface charges could attract the negative and positive charges from solution. Later, the subject of growing crystals will be considered again in terms of dislocations growing from imperfections in the lattice and these result when an atom or ion is missing in a particular plane.

A summary of a study of urea

In this section we are briefly going to look at the important stages in an X-ray diffraction study of urea. First, good crystals of urea were grown and were found to be tetragonal. Determination of the unit cell dimensions and density indicated that there were two molecules per unit cell. A study of the X-ray diffraction spectrum showed that there were absences in the diffraction spectrum which, though few, occurred for the (h00) plane when h was odd. But since $a = b$ in a tetragonal system then absences also occurred in the (0k0) plane when k is odd. The most probable space group was thought to be $P\bar{4}2_1m$ and a study of the positions likely for atoms in this particular space group is found by looking up the relevant data and are illustrated in Fig. 3.24. This leaves two solutions to the problem, one in which the hydrogen atoms are perpendicular to the planes containing the carbon, oxygen and nitrogen and one in which the molecule is in one plane. Urea is $H_2N.CO.NH_2$ with each nitrogen having a lone pair of electrons, two hydrogens and one carbon. The full arrangement around each N is trigonal unlike ammonia and the hydrogens are all in the plane.

Some of the equivalent positions for the space group $P\bar{4}2_1$ m

Multiplicity of positions	Point symmetry	Coordinates of equivalent positions
8	1	$x, y, z; \frac{1}{2} - x, \frac{1}{2} + y, \bar{z}$ $\bar{x}, \bar{y}, z; \frac{1}{2} + x, \frac{1}{2} - y, \bar{z}$ $\bar{y}, x, \bar{z}; \frac{1}{2} + y, \frac{1}{2} + x, z$ $y, \bar{x}, \bar{z}; \frac{1}{2} - y, \frac{1}{2} - x, z$
4	m	$x, \frac{1}{2} + x, z; \bar{x}, \frac{1}{2} - x, z$ $\frac{1}{2} + x, \bar{x}, \bar{z}; \frac{1}{2} - x, x, \bar{z}$
4	2	$0, 0, z; 0, 0, \bar{z}; \frac{1}{2}, \frac{1}{2}, 3; \frac{1}{2}, \frac{1}{2}, \bar{z}$
2	mm	$0, \frac{1}{2}, z; \frac{1}{2}, 0, \bar{z}$
2	$\bar{4}$	$0, 0, \frac{1}{2}; \frac{1}{2}, \frac{1}{2}, \frac{1}{2}$
2	$\bar{4}$	$0, 0, 0; \frac{1}{2}, \frac{1}{2}, 0$

Fig. 3.24. The X-ray diffraction study of urea

Fourier analysis

It is worth while to point out that this method is an extremely important aspect of crystallography. Fourier's theorem states that any periodic function can be made by superimposing (adding) sufficient sinusoidal waves which have the right amplitude and phase. We can indicate how such a procedure may be used for diatomic and triatomic molecules if we assume that the

76

molecules are linear and there is one molecule per 'unit cell'. The same type of procedure is used for two or three dimensions. The electron density distribution of molecules may be obtained by superimposing the appropriate sinusoidal waves which are themselves representative of electron distribution.

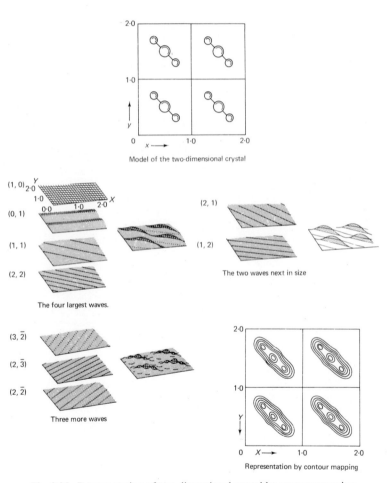

Fig. 3.25. Representation of two-dimensional crystal by contour mapping

Superposition may be done for an AB_2 molecule and in the figure there are four theoretical unit cells the electron density of which is obtained by superimposing nine sinusoidal waves of electron density. Note how the electron density of the molecule is gradually evolved from the gradual superposition of the waves.

If we just pause for a little while to think what is happening when X-rays are diffracted by crystals we shall see that Fourier analysis is needed to calculate the image. In the nineteenth century Ernst Abbé considered that the image formed by an ordinary microscope is formed in two stages. The object firstly scatters incident light in all directions forming a diffraction pattern and secondly the objective lens recombines the waves to form the image. The resolving power of the microscope depends on the aperture of the objective lens. The full X-ray structure analysis makes full use of the two stages except that the second stage is carried out with the aid of the Fourier analysis. In a normal experiment the rays are reflected from the crystal planes and form a series of diffraction spots of different intensity on a photographic film. The image of the electron density may be calculated because the X-rays are reflected by the electrons on the atoms to give the diffraction spots. One of the difficulties is however that the resultant intensity of the spots depends on the amplitude of the diffracted waves and is independent of the phase so that methods of determining the phase have to be derived. In the examples above if the phases were changed by half a period then all that would happen is that the crests would be replaced by troughs and the distribution of the image electron density would still resemble that of the molecules. There are therefore two possible positions of the waves in centrosymmetric structures and it is not really known whether to add or subtract these waves. For N waves there are 2^N possible ways of combining them so that without some means of determining the phases there would be some 2^{400} possible combinations for low resolution studies of biological molecules. Further when there is no centre of symmetry then there are no restrictions and there is an infinite number of ways in which the waves may be combined. We shall see that the problem can be solved by isomorphous substitution.

A full structural analysis is divided into five stages. First, the dimensions of the unit cell are determined using the Bragg technique. Secondly, the intensity of the spots is determined to give the amplitude of the waves. Thirdly, the appropriate phases of waves generated from these spots to form the image of the electron density are determined. The fourth step is to combine the waves from the diffraction spots by Fourier techniques to give the electron density map. Finally the crude image so obtained is improved or refined to give as true a picture of the molecule or structure as possible. The resolution must be as high as possible, that is the ability to distinguish between atoms or groups must be as high as possible. The electron density is calculated for a regular array of points in the image and the results are plotted on a contour of lines of electron density. The resolution of the final image is greater, if a large number of spots in the diffraction pattern are used. If the diffraction beam within a small angle is studied then the image has a low resolution which can be improved by using larger angles.

The full Fourier equation for the electron density is

$$\rho_{(xyz)} = \sum_h \sum_k \sum_l F_{hkl} \cos 2\pi \left[(hx + ky + lz) + \alpha_{hkl} \right]$$

$\rho_{(xyz)}$ is the electron density of the point whose coordinates are xyz.
F_{hkl} is the amplitude.
α_{hkl} is the phase.

hkl are the indices which may separately take values from $-\infty$ to $= +\infty$ but normally the contribution to the electron density decreases as the numerical values of the coefficients increase. The amplitude of the wave and the intensity of the diffraction spot are related as h_{hkl} is proportional to the square of F_{hkl}.

Questions

1 Define and explain the terms (a) electromagnetic spectrum; (b) X-ray diffraction; (c) interference; (d) diffraction to normal light.
2 Write down the advantages and disadvantages of the following methods of studying crystals using X-rays. (a) The Laue method; (b) rotation; (c) oscillation; (d) Weissenberg; (e) powder methods.
3 What do you understand, giving examples, by the terms:
  (a) Isomorphism; (b) polymorphism; (c) amorphism; (d) morphotropism?
4 Draw clear diagrams to illustrate or notes to explain (a) axis of symmetry; (b) plane of symmetry; (c) centre of symmetry; (d) screw axis; (e) glide plane; (f) space lattice; (g) unit cell; (h) primitive translation; (i) space groups; (j) urea has tetragonal crystals.
5 Explain the relationship between the classes of crystal, space lattices and space groups.
6 Describe in detail, giving theory where possible, how a full structural analysis of sodium chloride is carried out.
7 Illustrate how the following terms are important in crystallography: (a) density; (b) atoms or ions per unit cell; (c) wavelength of X-rays; (d) Bravais lattice.
8 Define: atoms; ions; molecules; elements; compounds.
9 Show how a monochromatic beam of X-rays is produced and how the wavelength is determined accurately.
10 To what extent may we regard X-rays as having a wave motion?
11 How does a knowledge of the external structure of a crystal help us to understand the internal structure of a crystal?
12 Discuss the work of (a) L'Abbé Haüy; (b) Mitscherlich; (c) Pasteur; (d) Le Bel; (e) van't Hoff; (f) Hiortdahl; (g) Bravais; (h) W. H. and W. L. Bragg; (i) Ernst Abbé.
13 Direct evidence of the wave nature of X-rays comes from the reflection of the rays from crystals. Discuss this statement mentioning Friedrich and Knipping, electromagnetic radiation, interatomic distances and wavelength, Bragg formula, diffraction patterns, Fourier analysis.

References

E.F.V.A., *X-ray Crystallography* (Advanced Science Series)

FARADAY SOCIETY, General Discussion, Crystal Structure and Chemical Constitution, 1929.

GLASSTONE, S., *Textbook of Physical Chemistry*, 2nd edn, Macmillan (1948), 1956, 340.

JAMES, R. W., *X-ray Crystallography*, 5th edn, Methuen, 1928.

LONSDALE, K. *Science News*, 1950, 59.

4   Ions and ionic crystals

In this and the following chapters we are going to consider different types of crystal according to whether the crystal is made up from ions or from molecules. We shall then consider metals, followed by the new and exciting studies of biological molecules. But what kind of information does an X-ray diffraction study of crystals give? In favourable cases accurate information is given about the relative positions of atoms, internuclear distances, co-ordination numbers, and structures of simple and complex molecules. The object of the remaining chapters is to develop gradually the factual knowledge of crystallography and where convenient it has been indicated how such information is derived practically.

Ionic crystals

A neutral atom consists of a heavy nucleus which contains protons and possibly neutrons. The atomic number of an element is defined as the number of electrons outside the nucleus or the number of protons inside the nucleus. The charge on each proton is $+1$ and, since the atom is electrically neutral, there is an equal number of electrons with a charge of -1. When one electron is removed to form an ion, for example the Na^+ ion, there is now one more positive charge than there are negative charges, the ion then is positively charged. The process of ionisation is therefore

$$M_{(gas)} \rightleftharpoons M^+_{(gas)} + e$$

Really the phase of the atoms and ions ought to be defined and it is usual to refer the ease with which an ion can be formed to the ionisation energy which we shall further discuss later in this chapter. For the moment, however, we will consider only the more simple aspects of the formation of ions. A negative ion may be formed by the addition of an electron to a neutral atom, for example

$$X_{(gas)} + e \rightleftharpoons X^-_{(gas)}$$

Such an ion could be the chloride ion, and in a solution of sodium chloride which is an ionic compound there are equal numbers of sodium ions and chloride ions. In a crystal of sodium chloride there are no molecules of sodium chloride but again just the ions. In a crystal the ions are localised but vibrate about a certain fixed point, and we can determine the position of the fixed point as we have seen by X-ray diffraction.

A circuit is set up with a battery and electrodes dipped into the electrolyte which contains ions of the solute. In a solution ions are more mobile and the positive ions are attracted toward the negative cathode and the positive ions therefore are called cations. The negative ions are attracted to the positive pole (anode) and are called anions. One fundamental rule is that like charges repel and unlike charges attract and so the sodium ions are attracted toward the negative chloride ions, whereas chloride ions will be repelled by similar chloride ions.

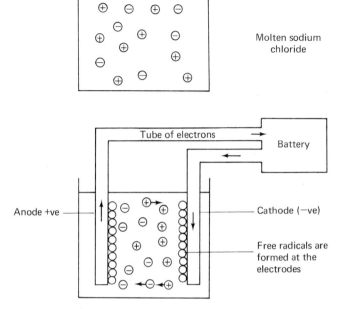

Molten sodium
chloride

Tube of electrons

Battery

Anode +ve

Cathode (−ve)

Free radicals are
formed at the
electrodes

Fig. 4.1. The electrolysis of sodium chloride

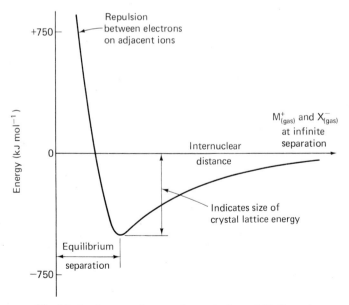

+750

Repulsion
between electrons
on adjacent ions

$M^+_{(gas)}$ and $X^-_{(gas)}$
at infinite
separation

Energy (kJ mol^{-1})

0

Internuclear
distance

Indicates size of
crystal lattice energy

Equilibrium
separation

−750

Fig. 4.2. Some energy changes when an ionic crystal is formed

If two ions M^+ and X^- are separated at infinity (Fig. 4.2) there is no large attractive force between the two sets of ions. As the ions are brought closer together they start attracting one another until, if sufficient ions of different kinds are present, the ionic structure may be formed. The depth of the well is a measure of the crystal lattice energy of the ionic compound. The energy is of the order of 750 kJ mol^{-1} of sodium chloride where 1 mole of sodium chloride is numerically equal to the molecular weight of sodium chloride expressed in grammes. The mole is the amount of substance which contains as many elementary units as there are atoms in 0·012 kilogramme of carbon-12. The elementary unit must be specified and may be an atom, a molecule, an ion, a radical, an electron, a photon, etc., or a specified group of such entities. The curve starts rising again because there is repulsion between the electrons on neighbouring atoms and this force is usually taken to be proportional to R^{-10}. Really one should not talk about the 'molecular weight' of sodium chloride but rather the 'ionic weight', as there are no molecules of sodium chloride, but unfortunately the term molecular weight is still applied to sodium chloride. Ionic crystals are formed as a direct result of this attractive force between opposite charges and the stability of a particular ionic compound is related to the lattice energy, where the lattice energy is the energy required for the process

$$M^+_{(gas)} + X^-_{(gas)} \rightleftharpoons M^+X^-_{(solid)}$$

The *crystal lattice energy* is defined as the heat liberated when 1 mole of the substance is formed from its constituent gaseous ions. The lattice energy is considered later in this chapter.

The major properties of ionic compounds are that in the molten state they conduct electricity and when dissolved in water they are electrolytes. Ionic compounds are insoluble in solvents such as benzene, which are covalent, but are often soluble in polar solvents like water.

In Fig. 4.3 the relationships between the types of bonds, that is ionic, covalent and metallic are illustrated. It must be emphasised that the boundary between them is by no means sharp or distinct, although ionic compounds, such as caesium fluoride, do not contain molecules, but some covalent compounds such as iodine do contain discrete molecules. Metals such as tungsten, aluminium or sodium are characterised by there being not one bond but a framework of metal ions through which electrons are allowed to move. Ionic bonds are non-directional, that is the charge is not pointed in a particular direction, so that in sodium chloride the sodium is surrounded by six chlorine ions and the chloride is surrounded by six sodium ions. The covalent bond is, however, directional and the electron wave motion tends to be more localised in iodine between the two iodine atoms and the strength arises due to the attraction between the nuclei and the electron wave motion. As previously intimated the metallic bond is not directional.

It is incorrect to say that quartz, for example, is ionised part of the time and covalent part of the time. Quartz, like many crystals, is in an intermediate state that is being partly ionic and partly covalent. The Si–O distance is *less* than either the sum of the ionic radii or of the sum of the covalent radii and the O–Si–O angle is also intermediate. Hydrides of some metals are intermediate between ionic covalent and metallic. In these interstitial hydrides the hydrogen particles pass into the holes in the structure of the metal which may

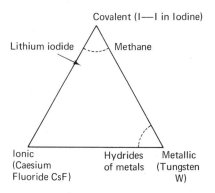

Fig. 4.3. The relationship between the bonds
The division between each type is not sharp

swell in order to hold the hydrogen. Hydrides may be ionic, metallic or covalent, for example sodium hydride (Na^+H^-) is ionic but methane (CH_4) is essentially covalent. Hydrides of many metals such as palladium hydride really consist of a framework of metal ions into which the hydrogen ions or hydrogen molecules are substituted. The distinction between ionic, covalent or metallic hydrides is clear but there is often no clear dividing line between ionic and covalent properties. Lithium iodide is intermediate between being an ionic and a covalent compound. In a completely ionic compound the electrons are entirely transferred from the cation to the anion but in lithium iodide the lithium still retains a partial share in the electron. The lithium cation has a small radius and consequently the charge is in a small volume which attracts or polarises the electron cloud around the iodide ion.

The structures of some ionic compounds

Let us have a close look at the structures in Fig. 4.4. Many structures of ionic covalent compounds and metals are based simply on the cubic structure ($a = b = c$, $\alpha = \beta = \gamma = 90°$). The cubic structure consists of an ion at each corner of a unit cell but in the body-centred cubic structure there is an additional ion in the middle of the cube. For example, caesium chloride has a simple cubic structure. The face centred structure consists of ions at each of the corners of the cube and in addition six ions at the centre of each face.

84

Fig. 4.4. The basic cubic structures

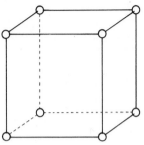

Atom or ion at each corner. Normal cubic
No example is known except possibly
α-polonium.

Additional ion in middle. Body-centred
cubic e.g. sodium, iron, tungsten.

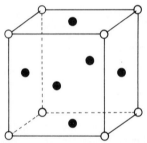

Six ions each at centre of face. Face-centred
cubic e.g., copper.

Fig. 4.5. An artist's impression of some cubic symmetry groups

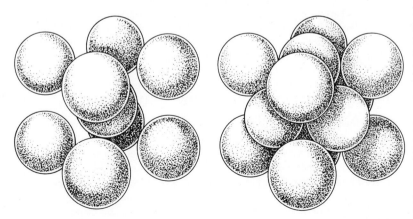

1. A body-centred cubic structure.

2. Part of a face-centre cubic structure.

An artist's impression of these groups of cubic symmetry (Fig. 4.5) illustrates the arrangement of the ions in the cubic cell. We shall be returning to these types of structures, particularly when we consider metals, but it should be noted that the lines that are drawn in the unit cell are only construction lines which join the centres of the ions in the unit cell, and do not represent bonds.

Caesium bromide

Like caesium chloride, caesium bromide has a simple cubic structure. Each caesium ion is surrounded by eight bromide ions and each bromide ion is surrounded by eight caesium ions. The ions are said to have a coordination number of eight because the ion is surrounded by eight ions which are (here) at the corners of a cube. Both the caesium ions and the bromide ions form simple cubic structures which interlock with each other. Four complete cubes of caesium ions are shown together with one face of the bromide ion cube and each ion forms the corner of six cubes. The external crystal structure faces discussed in Chapter 2 are usually closely related to the positions of the ions within the unit cell. The external faces could be parallel to the plane 5234 or 2165 which consist of alternate Cs^+ and Br^- ions. Cleavage also occurs parallel to such planes but not usually parallel to A2CD5B because there would be a resultant positive or negative charge at the surface.

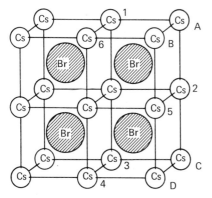

1. The simple cubic lattice

The caesium halides are not based on a simple body-centred lattice (BCC). They are based on a simple lattice with the caesiums at 000, X at $\frac{1}{2}\frac{1}{2}\frac{1}{2}$ the space group is Pm3m. Compare this with the sodium halides which are based on a face centred lattice (FCC) with the sodiums at 000, $0\frac{1}{2}\frac{1}{2}$, $\frac{1}{2}0\frac{1}{2}$, $\frac{1}{2}\frac{1}{2}0$ and the halides at $\frac{1}{2}\frac{1}{2}\frac{1}{2}$, $\frac{1}{2}00$, $0\frac{1}{2}0$, $00\frac{1}{2}$ and the space group symmetry is Fm3m

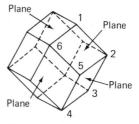

2. The crystal of a caesium bromide

Fig. 4.6. The crystal structure of caesium bromide (CsBr)

Sodium chloride

Sodium chloride belongs to the cubic crystal system and the crystals are usually cubic although they may be octahedral. The cleavage is perfectly cubic and the fracture is usually conchoidal in which the mineral has a curved fracture which is often concentric with the point at which the fracture started. In the crystal of sodium chloride it is incorrect to say that there are Na_6 or Cl_6 octahedra as each arrangement consists of six ions. Slightly more than one unit cell is illustrated in order to show how one ion is surrounded by six ions. For example, each sodium ion is surrounded by six chloride ions and each chloride ion is surrounded by six sodium ions. The arrangement of the six ions is said to be octahedral. The sodium ions form a face-centred structure and so do the chloride ions and the interlinking of the two structures give the complete ionic structure of sodium chloride.

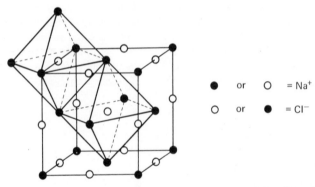

Fig. 4.7. The arrangement of octahedra within the structure of sodium chloride. Each ion of one kind is surrounded by six equidistant ions of the other kind

Sodium chloride and caesium chloride each are characteristic of two basic types of structures. For example, all the alkali halides (except caesium chloride, caesium bromide and caesium iodide) plus silver fluoride, silver calcium oxide have the sodium chloride structure. The caesium chloride structure is also typical of caesium bromide, caesium iodide, thallium chloride and thallium bromide. Surprisingly caesium chloride itself exhibits polymorphism because although caesium chloride crystallises at room temperature with the structure illustrated above, at the transition point 718 K its structure changes to that of sodium chloride. Compounds in which the structure changes at one particular temperature are said to be *enantiotropic* and many examples of such substances are known when the transition point is well above ordinary temperatures. The transition point of the two forms of zinc sulphide ZnS, wurtzite and zinc blende, is around 1300 K. Below this temperature wurtzite is more stable but above this temperature zinc blende or sphalerite is stable. It is therefore surprising that both forms are found at normal temperatures, but once formed and cooled, to change the structures requires a great deal of energy. *Monotropy* is said to occur where one substance is more stable under all temperatures at atmospheric pressure. Carbon is normally monotropic and as graphite is the stable form and diamond the unstable form (except at very high temperatures and pressures) it is surprising that diamonds exist at all. The reason for the existence of diamond at room temperature is that many solid state reactions are very slow indeed. Chemical reactions involve interaction and changes of ions or atoms; in the solid state these are essentially localised and the chemical reaction which normally would take place rapidly in the gaseous phase is limited to a very slow rate. The term *allotropy* is used to cover all cases where an element can exist in more than one crystalline or molecular form (but which have very similar chemical properties) for example, sulphur, phosphorus and carbon can separately exist in several different forms.

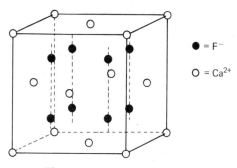

Fig. 4.8. Calcium fluoride (CaF_2)

Calcium(II) fluoride—fluorite

Calcium(II) fluoride belongs to the cubic crystal system and the crystals are usually cubic and more rarely octahedral or tetrahedral. Indeed the cleavage is perfectly parallel to the octahedron although the fracture can vary from conchoidal to uneven. The formula of calcium fluoride is CaF_2 (Fig. 4.8), that is for each calcium ion there are two fluoride ions. The *co-ordination number* of an ion is defined as the number of ions of opposite charge which are directly associated with that ion. In the case of sodium chloride the co-ordination number both of the sodium ions and the chloride ions is six. In calcium fluoride the formula (CaF_2) indicates that the coordination number of calcium must be twice that of the fluoride ions, and in the fluorite structure each calcium ion Ca^{2+} is surrounded by eight fluoride ions which are arranged at the corners of the cube, but each fluoride ion is surrounded by only four calcium ions at the corners of a tetrahedron. Compounds such as barium fluoride, strontium fluoride, have the fluorite structure but, as we saw in Chapter 3, magnesium fluoride has the rutile structure.

The fluorite structure may be obtained theoretically from the caesium chloride structure by removing alternate Ca^{2+} cations from each layer. The calcium ions present in reduced numbers must have double the charge of the fluoride ions and the value of a is 0·54 nm. This fluorite structure is one of the fundamental structures depending on the radius ratio effect for ionic compounds of the type AX_2. Thus the difluorides of metals with comparatively large radii have the fluorite structure and those fluorides of metals with small radii have the rutile structure (6:3) coordination.

Compound	M^{2+} radius 10^{-9} m	F^- radius 10^{-9} m	Ratio r^{2+}/r^-	Structure
CdF_2	·097	0·133	0·73	Fluorite
CaF_2	·099	0·133	0·75	Fluorite
HgF_2	0·110	0·133	0·83	Fluorite
SrF_2	0·112	0·133	0·85	Fluorite
PbF_2	0·120	0·133	0·9	Fluorite
BaF_2	0·134	0·133	1·0	Fluorite
MgF_2	0·066	0·133	0·5	Rutile
NiF_2	0·069	0·133	0·52	Rutile
CoF_2	0·072	0·133	0·54	Rutile
ZnF_2	0·074	0·133	0·56	Rutile
FeF_2	0·074	0·133	0·56	Rutile
MnF_2	0·080	0·133	0·6	Rutile

Titanium(IV) oxide (TiO_2)—Rutile

Titanium dioxide belongs to the tetragonal crystal system and the axial ratio is 0·644. The crystals often occur as tetragonal prisms which are terminated by pyramids. The crystals may twin on the second order (101) so that the two or more parts are at a sharp but definite angle and the twinning may be repeated until a wheel-shaped multiple twin results. Cleavage is not usually sharp but is roughly parallel to the (110) and (100) faces. The crystals have different colours according to the source but are usually reddish-brown owing to impurities.

In the rutile structure (Fig. 4.9) each metal ion (Ti^{4+}) is surrounded by six oxide ions (O^{2-}) as its direct neighbours and these ions are arranged at the corners of an octahedron. Each oxide ion has three Ti^{4+} neighbours at the corners of an equilateral triangle. The question to be answered shortly is 'Why should calcium fluoride have the fluorite structure and not the rutile structure?' Again, 'Why should a rutile be content with the rutile structure and not with the fluorite structure?'

The related crystal structures of $KLaF_4$ and $FeSbO_4$ are rather interesting, because in the first case potassium ions (K^+) and lanthanum (La^{3+}) ions have approximately the same size, and it is possible, therefore for potassium and lanthanum ions to occupy the same sites in the fluorite type unit cell. For this reason the fluoride ($KLaF_4$) has what is termed a random fluorite structure, that is the fluoride ions are arranged as in the fluorite structure but the potassium and lanthanide ions are arranged at random in the positions of the calcium ions. Again the iron and antimony ions are approximately the same size in the oxide and occupy at random the spaces in the rutile structure such that each is surrounded by six oxygen ions.

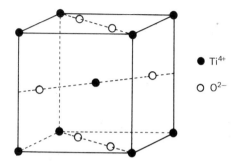

Fig. 4.9. Rutile: titanium(IV) oxide (TiO_2)

Tungsten(VI) oxide (WO_3)

In this structure tungsten (Fig. 4.10) is surrounded octahedrally by six oxygens and each oxygen is surrounded by two tungsten ions, but in the centre of the unit cell there is a space which may be filled by a univalent ion such as sodium. In the tungstates this space may be filled by the sodium ion to give $NaWO_3$. The oxidation number of the tungsten ion varies from six

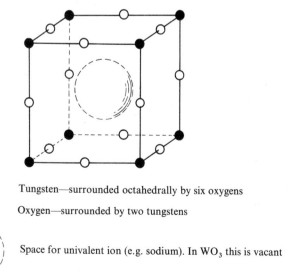

●   Tungsten—surrounded octahedrally by six oxygens

○   Oxygen—surrounded by two tungstens

     Space for univalent ion (e.g. sodium). In WO_3 this is vacant

Fig. 4.10. The cubic unit cell of tungsten(VI) oxide (WO_3)

1a	2a	3a	4a	5a	6a	7a	8			1b	2b	3b	4b	5b	6b	7b	0
H 1							Transition series										He 2
Li 3	Be 4											B 5	C 6	N 7	O 8	F 9	Ne 10
Na 11	Mg 12											Al 13	Si 14	P 15	S 16	Cl 17	Ar 18
K 19	Ca 20	Sc 21	Ti 22	V 23	Cr 24	Mn 25	Fe 26	Co 27	Ni 28	Cu 29	Zn 30	Ga 31	Ge 32	As 33	Se 34	Br 35	Kr 36
Rb 37	Sr 38	Y 39	Zr 40	Nb 41	Mo 42	Tc 43	Ru 44	Rh 45	Pd 46	Ag 47	Cd 48	In 49	Sn 50	Sb 51	Te 52	I 53	Xe 54
Cs 55	Ba 56	†La 57	Hf 72	Ta 73	W 74	Re 75	Os 76	Ir 77	Pt 78	Au 79	Hg 80	Tl 81	Pb 82	Bi 83	Po 84	At 85	Rn 86
Fr 87	Ra 88	‡Ac 89															

	4a	5a	6a	7a	8			1b	2b	3b	4b	5b	6b
†LANTHANIDES	Ce 58	Pr 59	Nd 60	Pm 61	Sm 62	Eu 63	Gd 64	Tb 65	Dy 66	Ho 67	Er 68	Tm 69	Yb 70	Lu 71
‡ACTINIDES	Th 90	Pa 91	U 92	Np 93	Pu 94	Am 95	Cm 96	Bk 97	Cf 98	Es 99	Fm 100	Md 101	No 102	Lw 103

Fig. 4.11. Periodic table

in WO_3 to five in $NaWO_3$. The *oxidation number* of an atom or ion is defined as 'the formal charge left on the ion when the attached groups are removed in their closed shell configuration'. For example, fluoride ions and chloride ions each like to have one negative charge and each oxygen ion has two negative charges. The removal of the three oxygen ions from tungsten (in WO_3) gives the tungsten an oxidation number of 3×2 or $+6$.

So far we have seen that there are different structures of ionic crystals but why do ions have different coordination numbers within the crystal and why are the structures not basically the same?

The attractive forces in an ionic crystal

Essentially, the main attractive forces within a crystal are the coulombic attractive forces. Within an ionic crystal, ions of opposite charge, wherever possible, will be next to one another, but ions of the same charge will be as far away from one another as possible. At first sight we might guess that somehow two factors will be important. The first would be the size of the ions and the second would be the charge on the ions themselves, that is, whether they carry a large or small positive or a negative charge. For example, we would expect that caesium ions would be larger than rubidium ions which in turn would be larger than potassium ions, sodium ions and lithium ions. A qualitative measure of the size of the ions would be in relation to the Periodic Table (4.11) which is a means by which elements are classified according to their atomic number. In the vertical series the names are in groups, from Group 1 up to Group 8. In the horizontal series these are given the periods of the elements, such as period 1 which consists of hydrogen and helium, period 2 consisting of lithium, beryllium, boron, carbon, nitrogen, oxygen, fluorine and neon.

Goldschmidt ionic radii

In a given group the atomic and ionic radii increase with increase in atomic number. The radii of many elements were determined by Goldschmidt from X-ray diffraction data and are listed in Fig. 4.12. Briefly, for a given element, where the positive charge of an ion increases the size of the ion decreases, because the positive nucleus is able to hold the electrons more closely, and therefore the calculated size of the Si^{4-} (0·198 nm) is greater than Si^{4+} (0·039 nm). The size of the ion increases with increasing negative charge, thus the size of the S^{2-} ion (0·174 nm) is larger than that of the S^{6+} ion (0·034 nm). As mentioned above the size of an ion increases for a given charge with increasing atomic number down a given group, thus the size of the fluoride ion (0·133 nm) is smaller than that of the iodide ion (0·229 nm). But how can one measure these ionic radii? If we just go back to the Bragg relationship that $n\lambda = 2d \sin \theta$ then d is the distance between

Symbol	Atomic Number	Ionic Species	Ionic Radius Å Empirical and Theoretical
H	1	H^{+}	
		H^{-}	1·27
Li	3	Li^{+}	0·78
Be	4	Be^{+2}	0·34
B	5	B^{+3}	0·2
C	6	C^{+4}	0·20
		C^{-4}	2·6
N	7	N^{+5}	0·1–0·2
		N^{-3}	1·71
O	8	O^{+6}	0·09
		O^{-2}	1·32
F	9	F^{+7}	0·07
		F^{-}	1·33
Na	11	Na^{+}	0·98
Mg	12	Mg^{+2}	0·78
Al	13	Al^{+3}	0·57
Si	14	Si^{+4}	0·39
		Si^{-4}	1·98
P	15	P^{+5}	0·3–0·4
		P^{-3}	2·21
S	16	S^{+6}	0·34
		S^{-2}	1·74
Cl	17	Cl^{+7}	0·26
		Cl^{-}	1·81
K	19	K^{+}	1·33
Ca	20	Ca^{+2}	1·06
Sc	21	Sc^{+3}	0·83
Ti	22	Ti^{+4}	0·64

Symbol	Atomic Number	Ionic Species	Ionic Radius, Å Empirical and Theoretical
V	23	V^{+4}	0·61
		V^{+5}	0·4
Cr	24	Cr^{+3}	0·65
		Cr^{+6}	0·34–0·4
Mn	25	Mn^{+2}	0·91
		Mn^{+4}	0·52
		Mn^{+7}	0·46
Fe	26	Fe^{+2}	0·83
		Fe^{+3}	0·67
Co	27	Co^{+2}	0·82
		Co^{+3}	0·65*
Ni	28	Ni^{+2}	0·78
Cu	29	Cu^{+}	0·96
Zn	30	Zn^{+2}	0·83
Ga	31	Ga^{+3}	0·62
Ge	32	Ge^{+4}	0·44
		Ge^{-4}	2·72
As	33	As^{+6}	0·47
		As^{-3}	2·22
Se	34	Se^{+5}	0·3–0·4
		Se^{-2}	1·91
Br	35	Br^{+7}	0·39
		Br^{-}	1·96
Kr	36		
Rb	37	Rb^{+}	1·49
Sr	38	Sr^{+2}	1·27
Y	39	Y^{+3}	1·06
Zr	40	Zr^{+4}	0·87
		Mo^{+4}	0·68

Symbol	Atomic Number	Ionic Species	Ionic Radius, Å Empirical and Theoretical
Ru	44	Ru^{+4}	0·65
Rh	45	Rh^{+2}	0·69
Pd	46	Pd^{+2}	0·50*
Ag	47	Ag^{+}	1·13
Cd	48	Cd^{+2}	1·03
In	49	In^{+3}	0·92
Sn	50	Sn^{+4}	0·74
		Sn^{-4}	2·94
Sb	51	Sb^{+5}	0·62
		Sb^{-3}	2·45
Te	52	Te^{+6}	0·56
		Te^{+4}	0·89
		Te^{-2}	2·21
I	53	I^{+7}	0·50
		I^{+5}	0·94
		I^{-}	2·29
Cs	55	Cs^{+}	1·65
Ba	56	Ba^{+2}	1·43
La	57	La^{+3}	1·22
Ce	58	Ce^{+4}	1·02
		Ce^{+3}	1·18
Pr	59	Pr^{+4}	1·00
		Pr^{+3}	1·16
Nd	60	Nd^{+3}	1·15
Sm	62	Sm^{+3}	1·13
Eu	63	Eu^{+3}	1·13
Gd	64	Gd^{+3}	1·11
Tb	65	Tb^{+3}	1·09

10Å = 1 nm

In Fourier analyses of quartz and of the silicates, say $ASiO_3$ or B_2SiO_4 the Si and the O are found to be of the same size. One cannot say therefore that in the above case of Si^{4+} that the radius is an empirical result (Si^{4+} are in fact hypothetical). The true silicates are not ionic, but are estimated to be approximately 50 per cent ionic only. Quartz would be regarded as $Si^{2+}[O^{-}]_2$ and not as $Si^{4+}O_2^{2-}$ or SiO_2.

Fig. 4.12. Crystallographic (ionic) radii

the layers off which the X-rays are reflected. If d can be related to the size of the ions then we can calculate the size of the ions. By choosing the right face the distance can be made to correspond to the distance between like ions and half this distance between the layers is the ionic radius. Thus Landé in 1920 found the distance between halide ions increases from fluoride to iodide and concluded that the ionic radii increased in the same order. If the layers of ions are assumed to be touching then half this distance between the layers is the ionic radius. Bragg (1927) found that the distance between the layers of oxide ions in silicates is (0·27 nm) so that if the ions are touching the radius of the oxide ion is (0·135 nm). Nowadays we usually take a value of (0·132 nm) for the oxide ion and (0·133 nm) for the fluoride ion and these values are the bases of the Goldschmidt ionic radii. The distance d can also be related to the distance between oppositely charged ions which is the sum of the two radii and knowing one value the other can be calculated. For example with the compound MgO the Mg–O distance is (0·21 nm) so that the calculated radius for Mg^{2+} is $0·21 - 0·132 = 0·078$ nm (0·78 Å).

The radius ratio effect

This applies for ionic compounds provided that the ions may be regarded as spheres. In some way the structure of a crystal is related to the size of the ions which constitute the unit cell, but really it is not the absolute size of the ions which is important but the ratio of the sizes. For example, in calcium fluoride the important factor is not the size of the calcium ions or the fluoride ions by themselves but rather the ratio of the size of the calcium ions to that of the fluoride ions. The rule is that ions with like charges must not touch and ions with unlike sign must touch if at all possible. The limiting ratios in Fig. 4.13 represent the ratios of the sizes of the cations (r_M) to those of the anions (r_X). For example, when an ion is surrounded by two others which are on opposite sides of the ion there is no limiting radius ratio as far as crystal structure is concerned, but when the coordination number is three and these three ions are arranged at the corners of an equilateral triangle (as the Ti^{4+} ions in the rutile structure surround the oxide ions) then the limiting ratio of the radius of the cation divided by that of the anion is 0·15. Similarly, other limiting ratios can be found. In Fig. 4.14 r_M is the radius of the cation M^{3+} and r_X is the radius of an anion X^-, the points A are at the vertices of the equilateral triangle. Then by geometry it is possible to calculate the ratio of r_M to that of r_X, and the reader is left to work this out for himself. The model set up for the MX_4 type of structure, that is with a coordination number 4 is rather more difficult to calculate and is a problem in three dimensional geometry, but the reader should be able to calculate the limiting radii for the coordination numbers 4 and 6, the model for which is illustrated in Fig. 4.15, and the reader will probably have more difficulty with the coordination numbers 6 and 8. (Fig. 4.16.)

Fig. 4.13. Limiting radius ratios

But as can be seen there are many exceptions to the rule when the ions cannot be regarded as hard spheres.

$$CsCl \text{ structure } 1\cdot0 \ < r_X/r_M < 1\cdot37$$
$$NaCl \text{ structure } 1\cdot37 < r_X/r_M < 2\cdot44$$
$$ZnS \text{ structure } 2\cdot44 < r_X/r_M < 4\cdot55$$

Examples

Caesium chloride structure		Sodium chloride structure		Zinc blende structure	
Salt	r_X/r_M	Salt	r_X/r_M	Salt	r_X/r_M
CsCl	1·1	NaCl	1·9	ZnS	2·1
CsBr	1·2	NaI	2·3	ZnSe	2·3
CsI	1·3	KCl	1·4	CuCl	1·9
TlCl	1·2	RbI	1·5	CuBr	2·0

Sometimes the limiting ratios are quoted r_M/r_X

Stereochemical arrangement	Coordination number	Radius ratio limits
Triangular	3	0·155 / 0·225
Tetrahedral	4	0·225 / 0·414
Octahedral	6	0·414 / 0·732
Cubic		

Before seeing how the radius ratio effect explains morphotropism it is important to realise that no mention has been made of whether or not crystal structures are really related to atomic weights. Indeed it is just the formal charge and the radius ratio of the ions which matters and not the atomic weights. The morphotropic problem has been then 'Why do the fluorides of calcium, strontium and barium have the fluorite structure but magnesium fluoride has the rutile structure?' Reference to the Goldschmidt radii for the ions concerned shows that the radius ratio for the first three fluorides falls within that of the fluorite structure, and that for magnesium fluoride falls within the range for rutile structure. This means that if magnesium fluoride had the fluorite structure it would be unstable because the fluoride ions would be touching one another, and such an arrange-

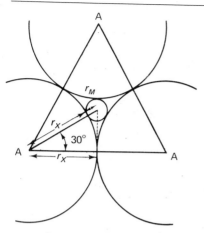

1. Coordination number 3 (MX$_3$)

r_M is the radius of cation M^{3+}.
r_X is the radius of anion X$^-$.
The points A are at the vertices of
an equilateral triangle.

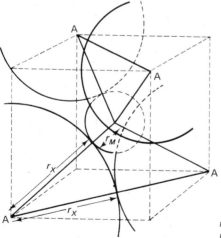

2. Coordination number 4 (MX$_4$) (tetrahedral)

r_M is the radius of cation M^{4+}.
r_X is the radius of the anion X$^-$.
The points A which are the nuclei
of the anions are at the vertices of a
tetrahedron.

Fig. 4.14. The radius ratio effect (of 'hard sphere' ions)

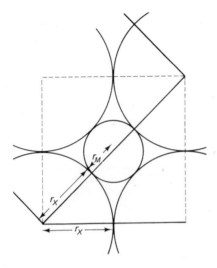

r_M = radius M^{4+} or M^{6+}.
r_X = radius of X^-.

r_M = radius M^{6+}.
r_X = radius of X^-.

3. Coordination number 4 MX_4 square
   Coordination number 6 MX_6 octahedral

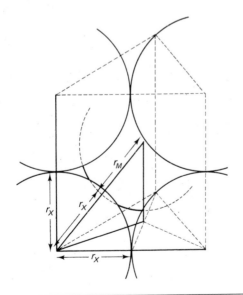

4. Coordination number
   6 MX_6 trigonal prism

Fig. 4.15. Radius ratio effect

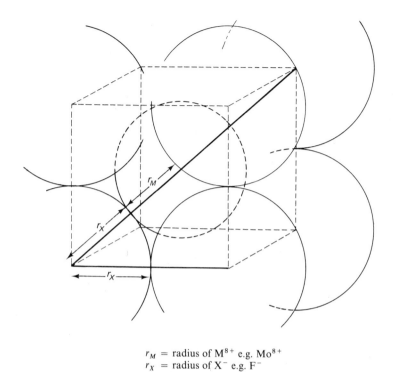

r_M = radius of M^{8+} e.g. Mo^{8+}
r_X = radius of X^- e.g. F^-

Fig. 4.16. Radius ratio effect coordination number 8, e.g. MoF_8 (cube)

ment is unstable. One limitation, however, of the radius ratio effect is that it can really be carried too far, the reason being that, as we have seen, ions such as iodide ions may be polarised, or the shape of the ion altered by the positive ions. The effect is that the true radius of the iodide ion is larger than that which one would expect when the ions are considered as hard spheres. The radius ratio effect is therefore limited to ionic compounds where the ions may be truly considered as hard spheres, and in particular when the nucleus is symmetrically surrounded by electrons in filled or half-filled shells.

Energy changes involved in the formation of the crystal of sodium chloride

Hess's Law states that the energy changes involved in a chemical re-action are independent of the paths taken. The Born–Haber Cycle uses this law in representing the formation of a crystal of sodium chloride by one of two ways (Fig. 4.17). First, by direct combination of the elements in

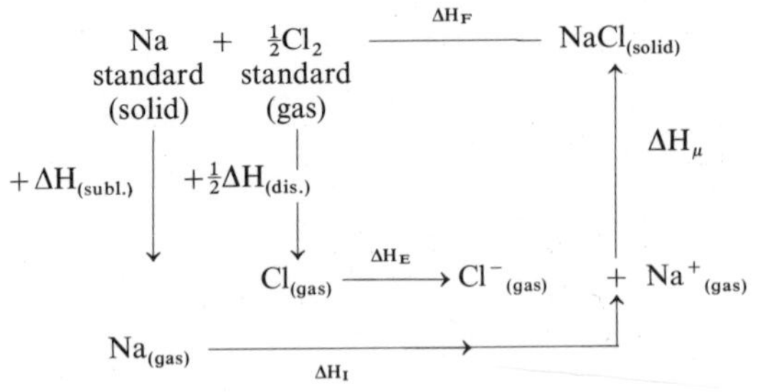

$$\Delta H_f = \Delta H_{(subl.)} + I + \tfrac{1}{2}\Delta H_{(dis.)} + A_{Cl} \pm \mu$$

Electron affinities and lattice energies are often negative.

Energy changes involve:

$$Na_{(s)} = Na_{(g)} \qquad\qquad \Delta H_{(subl.)}$$
$$Na_{(g)} = Na^+_{(g)} + electron\ (e^-) \quad \Delta H_I$$
$$\tfrac{1}{2}Cl_{2(g)} = 2Cl_{(g)} \qquad\qquad \tfrac{1}{2}\Delta H_{(dis.)}$$
$$Cl_{(g)} + e^- = Cl^-_{(g)} \qquad\qquad -\Delta H_E$$
$$Na^+_{(g)} + Cl^-_{(g)} = NaCl_{(s)} \qquad \pm\Delta H_\mu$$

$$Na_{(s)} + \tfrac{1}{2}Cl_{2(g)} = NaCl_{(s)} \qquad \Delta H_F$$

Fig. 4.17. The Born-Haber cycle for sodium chloride

their standard states, to give one mole of the crystal. The overall *heat of formation* ($\triangle H_f$) is defined as the heat evolved or used when one mole of the substance is formed from its constituent elements in their standard states, that is the phase solid, liquid or gas at 298 K and one atmosphere pressure, so that the standard state of sodium is solid and the standard state for chlorine is gaseous. The second and roundabout route involves first the sublimation of the metal to the gaseous phase to give sodium atoms and then the ionisation of the gaseous sodium atoms to give sodium ions. The energy evolved in the first process is *the Heat of Atomisation* ($\triangle H_A$) which is defined as the energy required to convert one mole of the element in its standard state to gaseous atoms. The energy for the second process is the first *ionisation energy* (kJ mol^{-1}) which is defined as the energy required to remove one electron from a neutral gaseous atom to give a gaseous ion. The gaseous sodium ions are able to combine with gaseous chloride ions to give the ionic crystal and the *lattice energy* ($\triangle H\mu$) is defined as the heat evolved when one mole of the substance at 298K is formed from its constituent gaseous ions. The first energy change for the chlorine molecule is to dissociate the bond. The *bond dissociation energy* ($\triangle H_B$) is defined as the energy required to liberate atoms from one mole of the gaseous substance. The *electron affinity* ($\triangle H_E$) is defined as the energy liberated when one electron is brought up to the neutral gaseous atom. We must limit our discussion of these heat changes but it must be emphasised that all changes take place under equilibrium conditions. The heat of formation may therefore be written as below—negative signs mean heat evolved and positive heat absorbed.

$$\pm F = +A + I + \tfrac{1}{2}B + E + \mu$$

(For simplicity $\triangle H$ signs have been omitted.)

The most important of these changes is really the balance of the ionisation energies and the lattice energies. The kind of problem we are now able to answer is 'Why should the formula for crystalline sodium chloride be NaCl and not $NaCl_2$?' The ionisation energies which are listed in Fig. 4.19 are determined by placing the neutral gaseous atoms in a discharge tube and applying a high voltage across the tube and measuring the potential at which the substance conducts electricity. Below this voltage the substance is composed of the neutral gaseous atoms but above this voltage the substance is essentially composed of the ions. The higher the ionisation energy the more difficult it is to form the ions, thus the first ionisation energy for sodium is low but the second ionisation energy for sodium in which the noble gas core is broken is extremely high and it is very difficult to form Na^{2+} ions. The second ionisation energy is the energy required to remove an electron from the gaseous ion which has a unit positive charge.

The lattice energies may be calculated using the Kapustinskii formula, which is illustrated in Fig. 4.18. The lattice energy of the halides of sodium may be calculated in the following manner:

$$\Delta H_{lattice} = -1200 \frac{\sum n . z_1 z_2}{r^+ + r^-}\left(1 - \frac{0.345}{r^+ + r^-}\right)$$

$\sum n$ = total number of ions, e.g., 3 in $MgCl_2$.

z_1, z_2 are the charges, e.g., 2,1 in $MgCl_2$.

$r^+ . r^-$ are the cationic and anionic radii respectively in Å.

In quick calculations the term in the bracket may be taken as approximately 0.9. ΔH in kJ mol^{-1}.

Fig. 4.18. Kapustinskii formula

For NaF $\quad \mu = \dfrac{-1200.2.1.1}{0.98 \times 1.33}\left(1 - \dfrac{0.347}{0.98 + 1.33}\right) \simeq -1700$ kJ

For NaI $\quad \mu = \dfrac{-1200.2.1.1}{0.98 \times 2.2}\left(1 - \dfrac{0.347}{0.98 + 2.2}\right) \simeq -1000$ kJ

In general the lattice energy decreases from the fluoride to the iodide for any given metal and hence fluorides are often much more stable than the iodides. It should be clear, however, that all of the above lattice energies are large and negative (stable) and hence all the sodium halides are stable. The heats of formation may be found in the usual way:

1a	2a	3a	4a	5a	6a	7a	8	8	8	1b	2b	3b	4b	5b	6b	7b	0
H 1 13·6																	He 2 24·6
Li 3 5·4	Be 4 9·3											B 5 8·3	C 6 11·3	N 7 14·5	O 8 13·4	F 9 17·4	Ne 10 21·6
Na 11 5·1	Mg 12 7·6											Al 13 6·0	Si 14 8·1	P 15 11·0	S 16 10·4	Cl 17 13·0	Ar 18 15·8
K 19 4·3	Ca 20 6·1	Sc 21 6·6	Ti 22 6·8	V 23 6·7	Cr 24 6·8	Mn 25 7·4	Fe 26 7·9	Co 27 7·9	Ni 28 7·6	Cu 29 7·7	Zn 30 9·4	Ga 31 6·0	Ge 32 8·1	As 33 10·0	Se 34 9·8	Br 35 11·8	Kr 36 14·0
Rb 37 4·2	Sr 38 5·7	Y 39 6·6	Zr 40 7·0	Nb 41 6·8	Mo 42 7·2	Tc 43 7·4	Ru 44 7·5	Rh 45 7·7	Pd 46 8·3	Ag 47 7·6	Cd 48 9·0	In 49 5·8	Sn 50 7·3	Sb 51 8·6	Te 52 9·0	I 53 10·4	Xe 54 12·1
Cs 55 3·9	Ba 56 5·2	†La 57 5·6	Hf 72 5·5	Ta 73 6·0	W 74 8·0	Re 75 7·9	Os 76 8·7	Ir 77 9·2	Pt 78 9·0	Au 79 9·2	Hg 80 10·4	Tl 81 6·1	Pb 82 7·4	Bi 83 8·0	Po 84 ≈8·2	At 85 ≈9·0	Rn 86 10·7
Fr 87 3·7	Ra 88 5·3	‡Ac 89 5·5															

Transition series

†LANTHANIDES

	4a	5a	6a	7a	8	8	8	1b	2b	3b	4b	5b	6b	7b
	Ce 58 5·6	Pr 59 6·9	Nd 60 6·3	Pm 61 5·8	Sm 62 5·6	En 63 5·7	Gd 64 6·2	Tb 65 6·7	Dy 66 6·8	Ho 67 ×6·5	Er 68 ≈6·5	Tm 69 ≈6·5	Yb 70 6·2	Lu 71 5·0

eV × 96·5 gives kJ mol^{-1}

Fig. 4.19. Ionisation energies (eV) of the elements

$$\pm F = I + A + \tfrac{1}{2}D + E + \mu$$

$$NaF = 504 + 209 + 153 - 348 - 1700 \simeq -1200 \text{ kJ mol}^{-1}$$
$$NaCl = 504 + 209 + 121 - 362 - 1200 \simeq -700 \text{ kJ mol}^{-1}$$
$$NaBr = 504 + 209 + 196 - 338 - 1100 \simeq -500 \text{ kJ mol}^{-1}$$
$$NaI = 504 + 209 + 152 - 309 - 1000 \simeq -450 \text{ kJ mol}^{-1}$$

The second ionisation energy for sodium is 4580 kJ and the lattice energy for NaF_2 is calculated by:

$$\mu = \frac{-1200.3.1.2 \times 0.9}{0.78 \times 1.33} \simeq -5600 \text{ kJ}$$

where r_M is estimated by noticing that the radius of Mg^{2+} is 0·078 nm (0·78 Å) and we can take this value to be slightly smaller than the Na^+ radius. In practice it is best to take a larger value of the radius where there is doubt because the larger value will give a smaller value of the lattice energy.

$$\pm F = I_1 + I_2 + A + D + 2E + \mu$$
$$\simeq 500 + 4600 + 200 + 300 - 700 - 5600 \simeq -700 \text{ kJ mol}^{-1}$$

That is NaF_2 is unstable with respect to NaF.

The lattice energy is related to the radius ratio for particular compounds (Fig. 4.20). Thus the lattice energy of caesium chloride for which the radius ratio is (r_M/r_X) is 0·75 higher than that of sodium chloride in which the radius ratio is 0·43 and zinc sulphide in which the radius ratio is 0·22. The absolute values for the coulombic lattice energies, however, have been omitted, the approximate values being 400 kJ mol^{-1}, but the reader is recommended to work out the values of these, using the Kapustinskii relationship. All the radii required are listed in this chapter. The coulombic lattice energies of ZnS and NaCl are so close that polarisation effects determine which structure is taken (see Chapter 5).

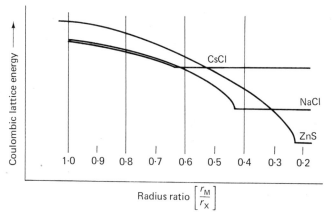

Fig. 4.20. The lattice energy is related to the radius ratio

102

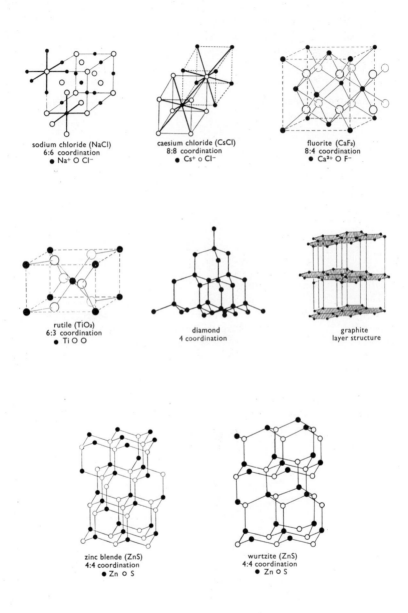

sodium chloride (NaCl)
6:6 coordination
● Na⁺ O Cl⁻

caesium chloride (CsCl)
8:8 coordination
● Cs⁺ o Cl⁻

fluorite (CaF₂)
8:4 coordination
● Ca²⁺ O F⁻

rutile (TiO₂)
6:3 coordination
● Ti O O

diamond
4 coordination

graphite
layer structure

zinc blende (ZnS)
4:4 coordination
● Zn O S

wurtzite (ZnS)
4:4 coordination
● Zn O S

Fig. 4.21. Some crystal forms

103

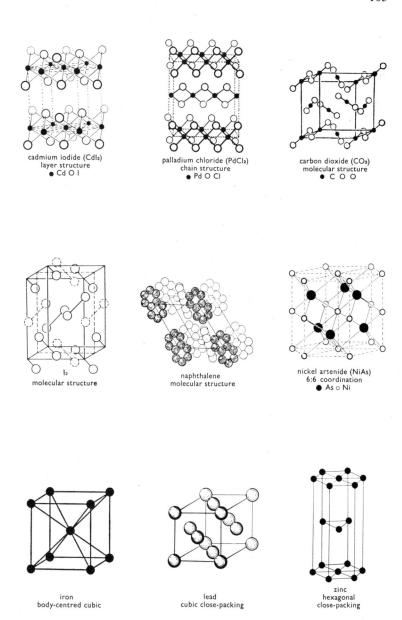

cadmium iodide (CdI₂)
layer structure
● Cd ○ I

palladium chloride (PdCl₂)
chain structure
● Pd ○ Cl

carbon dioxide (CO₂)
molecular structure
● C ○ O

I₂
molecular structure

naphthalene
molecular structure

nickel arsenide (NiAs)
6:6 coordination
● As ○ Ni

iron
body-centred cubic

lead
cubic close-packing

zinc
hexagonal
close-packing

Fig. 4.21 (cont.)

104

Questions

1 Draw a diagram to illustrate the direction of the electron flow in the electrolysis of calcium fluoride.

2 List as many different types of ionic and covalent structures as you can and state how the positions of the elements involved are related to the Periodic Table.

3 Draw clearly the structures of barium chloride and magnesium fluoride.

4 Calculate the limiting radius ratios from Figs. 4.14, 4.15 and 4.16.

5 Calculate the lattice energies of caesium chloride and zinc sulphide utilising the Kapustinskii relationship in Fig. 4.19.

6 Draw clearly, from memory, the structures of caesium bromide, sodium chloride, calcium chloride, titanium(IV) oxide and tungsten(IV) oxide.

7 Write down the limiting radius ratios for ions of coordination numbers 2, 3, 4, tetrahedral, 4 square and 6 and 8.

8 What do you understand by the term 'Goldschmidt radii'? To what extent do the absolute values of these radii govern the structure of crystals of the types M^+X^-, $M_2^+2X^-$ and $M_3^+3X^-$?

9 Why is the formula for crystalline caesium chloride CsCl and not $CsCl_2$?

10 Why is the formula for crystalline magnesium chloride $MgCl_2$ and not MgCl or $MgCl_3$?

11 List as many covalent molecules, with their structures, as possible.

12 Draw out the structures of the following from memory NaCl, CsCl, CaF_2, TiO_2, diamond, graphite, zinc blende, wurtzite, CdI_2, $PdCl_2$, CO_2, I_2, naphthalene, NiAs, Fe (one form only), Pb, Zn.

13 Describe the relationship between the ionic, covalent and metallic structures.

14 What is meant by a basic cubic structure and what are the basic ionic structures?

15 Show how crystallography has led to an explanation of the external appearance and cleavage of crystals.

16 What evidence does crystallography give regarding the existence of NaCl molecules. How do Goldschmidt radii aid your argument?

17 How does a knowledge of ionic radii help us to determine the possible coordination numbers in compounds of the general type MX, MX_2, MX_3, MX_4, MX_6?

18 If atoms are vibrating how is it that an ionic radius can be described?

References

GRAY, H. B. and HAIGHT, G. P., *The Basic Principles of Chemistry*, Wiley, 1966.

HUDSON, M. J., *Energy and Bonding*, English Universities Press, 1969.

MOFFATT, W. G., PEARSALL, G. W. and WULFF, J. *The Structure and Properties of Materials. Vol. 1: Structure*, Wiley, 1967.

5   Crystals with some covalent character

In the last chapter (see Fig. 4.21) we considered the simple ionic structures. Now we shall see how the ideas developed there can be extended, first to crystals which contain not just simple ions but the more complex ions, such as carbonate ions, then the formation of layer structures, macromolecular crystals, framework structures will be discussed, together with Fajan's rules and covalent radii.

In an ionic bond there is a complete transfer of an electron from the anion to the cation. In a covalent bond however classical theory normally regarded that the two or more atoms involved in the bond each donate one electron to that bond. Therefore in the covalent bond between two hydrogen atoms in the hydrogen molecule each hydrogen atom donates one electron to that bond (Fig. 5.1). There is not a complete transfer of an electron when a covalent bond is formed. In Chapter 4 the Periodic Table was referred to and although we cannot go into this in great detail the type of bond formed is often influenced by the structure of the nearest noble gas. For example, sodium gained the noble gas structure by losing one electron, and chlorine in sodium chloride gained the noble gas structure by gaining one electron. However, in the chlorine molecule both chlorine atoms can achieve the noble gas structure by sharing two electrons, and similarly, the two hydrogen atoms in the hydrogen molecule can gain the noble gas structure of helium by sharing two electrons. The noble gas structure can be achieved by carbon in carbon tetrachloride by sharing four electrons with each of the four

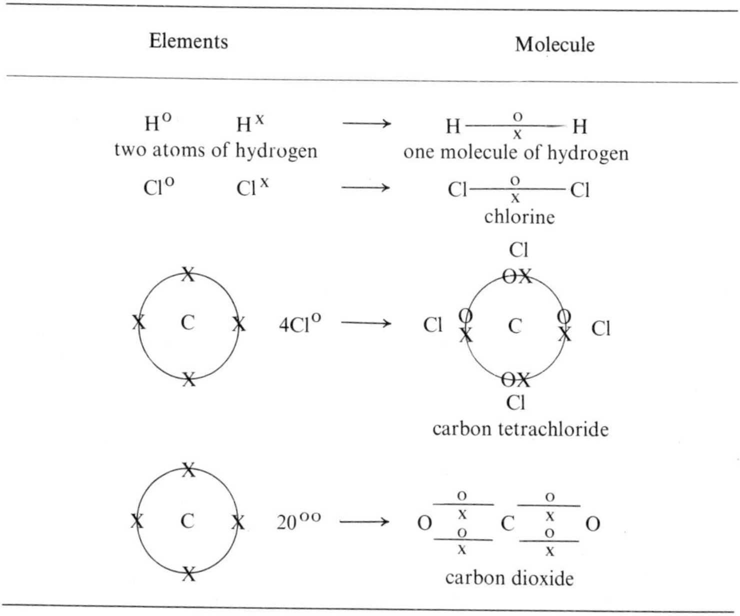

Elements	Molecule

* O, X represent electrons.

Fig. 5.1.

chlorines. The chlorines in carbon tetrachloride gain the noble gas structure in the same way that they did in the chlorine molecule.

In carbon dioxide (Fig. 4.21), oxygen needs two electrons to gain the noble gas structure of neon, and carbon needs four electrons to gain the same noble gas structure. Both atoms can gain the noble gas structure by sharing two pairs of electrons in forming a double bond between the carbon and oxygen. In diamond each carbon atom is bonded to four carbon atoms by four single bonds, all the carbon atoms can achieve the noble gas structure by forming single bonds.

In the series of compounds ethane C_2H_6, ethene C_2H_4 and ethyne C_2H_2, it can be shown either by electron diffraction or by X-ray diffraction that the carbon atoms are joined to one another. Modern evidence shows that the C—C bond length decreases from 0·154 nm (1·54 Å) in ethane to 0·135 nm (1·35 Å) in ethene and to 0·12 nm (1·2 Å) in ethyne. In some way these bond lengths must be related to the bond order (number of bonds) between the carbon atoms. The bond order is defined as the number of pairs of electrons which are shared in a covalent bond. For example, the bond order in ethene is two (Fig. 5.2). Each carbon atom in ethane is attached to three hydrogen atoms and one carbon atom and if the carbon atom shares one pair of electrons with each of these atoms then the noble gas structure of neon is obtained. In ethene, however, the carbon is attached to only two hydrogen atoms and one carbon atom and if it shared electrons with each of these three atoms the carbon atom would still be short of the neon structure by one electron. Each carbon atom can gain the neon structure by forming a bond of order two, that is by sharing an additional pair of

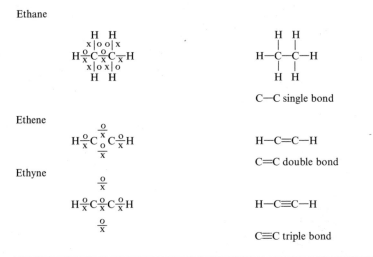

Fig. 5.2.

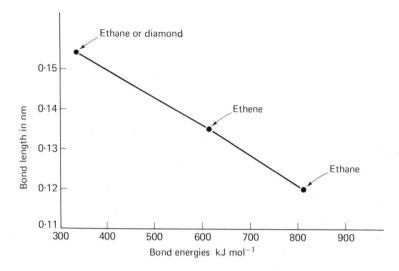

Fig. 5.3. The bond lengths and bond strengths of some carbon to carbon bonds

electrons with each other, and the bond between the two carbon atoms is referred to as a double bond. Similarly in ethyne the carbon atom is attached to one hydrogen atom and one carbon atom and the noble gas structure can be gained by forming a triple bond or a bond of order three between the two carbon atoms. In Fig. 5.3 bond order is plotted against bond length. The bond length decreases with increasing bond order. Qualitatively one would expect that the bond in ethyne would be stronger than that in ethane, and in fact the bond in ethyne is stronger than that in ethane but not three times as strong. The bond strength in ethane 347 kJ mol^{-1} and the bond strength in ethyne is 840 kJ mol^{-1}.

Covalent compounds are characterised by the fact that they do not conduct electricity, they often have low melting and boiling points because in covalent compounds there are discrete molecules, such as in benzene C_6H_6 there are individual molecules of benzene. The forces between the molecules, the intermolecular van der Waal's forces, are much weaker than the covalent bond. An average value for the intermolecular forces is 20 kJ mol^{-1} compared with 400 kJ mol^{-1} for the covalent bond. In rough terms it is twenty times easier to separate individual molecules than to break up one covalent bond. The result of the differences in bond strength are that in covalent crystals the molecules tend to exist as separate species. Thus a covalent crystal contains individual molecules which are often the same shape as in the gas phase. Thus in molecular structures of carbon dioxide and iodine there are separate molecules of each substances, and such structures are called molecular crystals. Although there is not usually a change in structure from the gaseous phase to the solid phase, with PCl_5 there is a dramatic change from PCl_5 molecules in the vapour to the ions $(PCl_6)^-$ and

$(PCl_4)^+$ in the solid, when the phosphorus partially achieves the noble gas configuration. Layer structures are often formed when the electrons on the iodide or sulphide ions are strongly attracted to the metal (see Fajan's rules) and cadmium iodide is a typical example, like palladium(II) chloride.

Again in solid carbon dioxide or naphthalene there are discrete molecules. In ionic compounds, however, in which there are ionic bonds there are no discrete molecules. It is not always true to say that covalent compounds consist of small molecules, because diamond consists of interlinking carbon chains and one diamond crystal is one complete molecule, and so if you are asked whether you have seen a molecule before, if you have seen diamonds, or even silica sand, then you have seen one complete molecule. Covalent compounds do not conduct electricity because there are no free anions or cations or residual charges which are able to move around and conduct the electricity. Sodium chloride does not conduct electricity when solid because the ions are unable to move about as they are fixed in the ionic structure. When sodium chloride is molten, however, the ions are able to move about so that many molten ionic compounds conduct electricity. The molten covalent compounds cannot conduct electricity because the individual molecules do not carry charges although they are able to move about. In metals the positive ions are fixed and the valency electrons are free to move through the whole structure. Valency electrons are those which the metal uses in chemical reactions so that each sodium, say, ion has one electron and each magnesium ion has two electrons.

Structures of some compounds

Calcium(II) Carbonate (Calcite)

Calcium carbonate belongs to the trigonal system and there are three main habits which are nail-head, dog tooth and prismatic. For a further discussion of such twins the reader should read F. Rutley's *Elements of Mineralogy*, which has been brought up to date by Professor H. H. Read (Allen and Unwin, 1963). The cleavage is perfectly parallel to the unit rhombohedron and the conchoidal fracture is often difficult to see because of the perfect cleavage.

Calcite consists of ions of Ca^{2+} and CO_3^{2-} anions. There is a covalent bond between the oxygen and carbon in the carbonate anion but it is not a simple covalent bond of order one. In Fig. 5.4 the structure of the carbonate anion is illustrated. Each oxygen forms one single bond with the carbon and one forms a double bond with the carbon, and this leaves a residue of two negative charges over the whole anion, but really there is no way of distinguishing between the three oxygens and we could equally well have drawn the double bond between any of the three oxygen atoms and the carbon atom. In fact the true structure of the carbonate ion can be found by theoretically combining the three structures. It must be emphasised that these structures, which are called canonical structures, are entirely theoretical and the true structure is a combination or hybrid of the three.

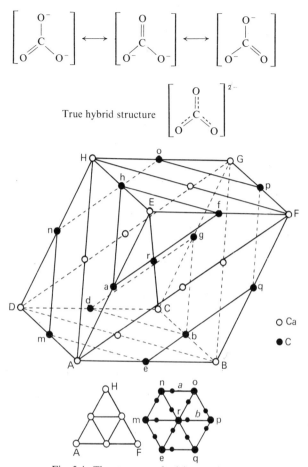

Fig. 5.4. The structure of calcium carbonate

The theoretical process by which these structures are combined is called *resonance*. Again resonance here is an entirely theoretical process. The bond order between the carbon and the oxygens is 1·33. The crystal of calcite is, therefore, made up of calcium ions and carbonate ions. The carbonate ions are planar, the three oxygen atoms surround the carbon atom at the corners of an equilateral triangle with the carbon atom at the centre of the triangle. Plate 5.1 shows a model of the calcite structure in two directions. First, in one direction there are not really many molecules, but in a direction at right angles to this structure there are a great many molecules and so the refractive index is different in these two directions. Calcite exhibits the property of birefringence which is dealt with very clearly in many advanced level books. Birefringence, as the name implies, is when a double image is seen when a single object is viewed through a crystal (of calcite). The double image arises because the light is refracted in different manners

110

depending on the refractive index of calcite in the two mutually perpendicular directions. Light is two wave motions which are mutually at right angles, so that Snell's Law applies to the two parts of light.

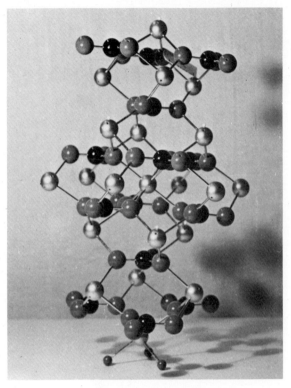

Plate 5.1. Calcite $(CaCO_3)$

In the elucidation of the structure of calcite the important measurements made were the dimensions of the unit cell and the positions of the oxygen atoms because by fixing the position of the oxygen atoms the position of the carbon atom is at the centre of the triangle of oxygen atoms. This arrangement of the carbonate anion is part of the requirement for a trigonal symmetry. The position of the calcium ions could be found by close attention to the intensity of the diffraction spots. The calcite structure which is also the typical structure of many ABO_3 compounds is formed when the atom A is relatively small having a co-ordination number of six, as in the compounds $LiNO_3$, $NaNO_3$, $MgNO_3$, $ZnCO_3$, $CdCO_3$, $ScBO_3$ and YBO_3. When A is large the aragonite structure is formed and the coordination number of A is increased definitely to nine and possibly to twelve in $RbNO_3$ and $CsNO_3$. The aragonite structure is characteristic of some compounds of the type ABO_3 such as the compounds KNO_3, $SrCO_3$, $BaCO_3$, $LaBO_3$, $RbNO_3$ and $CsNO_3$.

Diamond and graphite

Diamonds belong to the cubic system and the crystals are usually octahedral and cleavage occurs perfectly parallel to the faces of the octahedron so that diamond cutting means the art of tapping a crystal in the right place to make the crystal cleave as desired. Diamonds often have a curved face and are usually twinned. Graphite belongs to the hexagonal system and does not usually form crystals being usually flaky and cleavage occurs parallel to the large surface of the crystals. The crystals appear to be metallic and graphite is unusual in that it is non-metallic yet it conducts electricity.

Diamond and graphite (Plates 1.1 and 1.2) are allotropes of carbon yet surprisingly their properties are very different. Diamond as we have seen from Moh's hardness scale is one of the hardest elements known and yet graphite is a lubricant. As we have seen in Chapter 1 the difference in properties is closely related to the crystal structure of the two allotropes. The structure of diamond may be elucidated in the following way. The diamond crystal was mounted on the axis of rotation of the spectrometer and three photographs were obtained when the axis of rotation coincided respectively with the cube edge of the crystal, the cube diagonal and the face diagonal. At a first examination the structure was shown to be that of a face-centred cubic crystal. Let a be the edge length of a cubic cell. The density of diamond is 3.51 g/cm^3 and the mass of the carbon atom is $12 \times 1.64 \times 10^{-24}$ g. If there are n carbon atoms in the face-centred cubic cell, then

$$a^3 \times 3.51 = n \times 12 \times 1.64 \times 10^{-24}$$

from which $a = 1.776 \times n^{0.33}$ atomic units

By subsequent experiments it can be shown that $n = 8$ and therefore

$$a = 0.3552 \text{ nm or } 3.552 \text{ Å}$$

The diamond structure consists of one carbon atom which is surrounded tetrahedrally by four other carbon atoms. The structure is sometimes drawn as a face-centred cube represented by four such carbon atoms, which are displaced tetrahedrally. The unit cell of diamond is *either* the face-centred cubic cell which contains eight carbon atoms at

$$000 \quad 0\tfrac{1}{2}\tfrac{1}{2} \quad \tfrac{1}{2}0\tfrac{1}{2} \quad \tfrac{1}{2}\tfrac{1}{2}0 \quad \tfrac{111}{444} \quad \tfrac{133}{444} \quad \tfrac{313}{444} \quad \tfrac{331}{444}$$

or the simple $60°$ rhombohedral cell containing two carbon atoms which is one-quarter of the above volume. It is incorrect to say that the unit cell of diamond consists of one carbon atom which is surrounded tetrahedrally by four others. To say that four of these units are linked together to form a face centred cell is a description of the structure but not of the unit cell. All carbon atoms therefore have a valency of four, and have the noble gas structure of neon, and diamond is therefore hard and chemically inert, with the distance between the carbon atoms in diamond 0.154 nm. In graphite, however (Fig. 5.5), there are two main distances, one of 0.142 nm and the other 0·341 nm and the crystal of graphite is hexagonal consisting of separate sheets of carbon atoms in which there is a hexagonal framework of carbon atoms. In each sheet the distance between the carbon atoms is 0·142 nm and comparison of this distance with Fig. 5.3 shows that the bond is intermediate between a single and a double bond, but the very large distance 0·341 nm is the distance between the layers of carbon atoms. Since the distance is so long the bond must necessarily be very weak and although, at the time, early crystallographers were puzzled by this long bond it is now recognised that it is a typical example of a bond called a van der Waal's bond. These are very weak bonds and occur between molecules and are called intermolecular bonds. The energy of van der Waal's bond is approximately 20 kJ mol^{-1} compared with a value of 400 kJ mol^{-1} for each of the covalent and ionic bonds. The difference in structures between graphite and diamond therefore makes it easy to explain the difference in chemical and physical properties of the two allotropes. Graphite is a flaky substance and, when beaten in a mortar, breaks down into smaller and smaller flakes. The lubricating properties of graphite arise because the layers are able to move over one another and the great distance between the layers means that they can do so fairly easily.

The structures of diamond and zinc blende are very simple and the structures were elucidated by reference to a few photographs taken along the appropriate axes. In more complex structures, for example in graphite, it is necessary to look at the intensity of the diffraction spots. The intensity depends first on the directions along which the photographs are taken, and secondly, the intensity depends on imperfections in the crystal structure. The dependence of intensity with the orientation of the photograph will be discussed briefly now in connection with zinc sulphide, and imperfections are discussed in Chapter 6.

Zinc blende, zinc sulphide (ZnS)

Zinc blende (sphalerite, black jack) belongs to the cubic system. The crystals are usually tetrahedral or rhombododecahedral but are often twinned into unusual shapes although the cleavage is perfectly parallel to the faces of the rhombododecahedron. The fracture is usually conchoidal. A preliminary examination shows that the lattice is face-centred cubic and that there are four molecules in the face-centred lattice which may be drawn (Fig. 5.5) by placing zinc atoms at the corners of the cube and at the centre of each face. This is a very highly symmetrical arrangement of the zinc atoms and the sulphur atoms must be inserted with the observation that a crystal of zinc sulphide is polar along each of its diagonals. The only way to explain this is to place the sulphide atoms in planes parallel to the cube diagonals. The structure is the same as that of diamond except that half the atoms are zinc and half the atoms are sulphur atoms. Every zinc atom is surrounded tetrahedrally by four sulphur atoms and each sulphur atom is surrounded tetrahedrally by four zinc atoms. The photographs along different axes show spots of different intensities for the following reasons. For

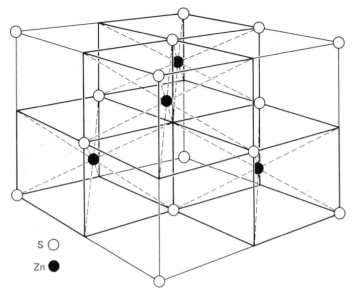

Separation of zinc and sulphur layers

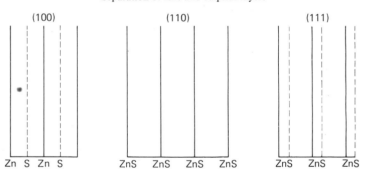

Fig. 5.5(a). Structure of zinc blende ZnS

example, reflection from the (100) plane gives low intensity spots because there are alternate planes of zinc and sulphur atoms. The reflection from the (111) face is more intense than that in diamond because the zinc and sulphur atoms reflect X-rays in different ways. The intensity of X-rays is dependent upon the atomic number of the atom because X-rays are really reflected by the electrons outside the nucleus. For this reason X-rays are reflected more from heavy atoms than from light atoms, but there is no real problem in zinc sulphide because both the zinc and the sulphur atoms are fairly heavy. The maximum intensity spots are from the (110) plane because these do not consist of alternate planes of zinc and sulphur units.

Wurtzite

In wurtzite, which is the second structure of zinc sulphide, both the sulphur atoms and the zinc atoms are tetrahedrally surrounded by four opposite atoms. In wurtzite, however, the sulphur atoms occupy hexagonal close packed positions and not the cubic close packed positions as in zinc blende. The two structures are related by moving parts of the structure parallel to the planes of the close packed atoms (see Fig. 5.5).

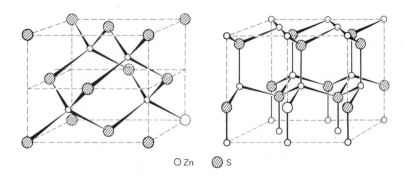

O Zn   ⊘ S

Fig. 5.5(*b*). Comparison of the two structures of zinc(II) sulphide

One of the forms of ice possesses the wurtzite structure and between each molecule there is hydrogen bonding so that each oxygen is surrounded tetrahedrally by four hydrogen atoms with two lying much closer than the other two. Interestingly the sizes of the unit cells of ice and silver(I) iodide are very similar so that the iodide has been used as a nucleus around which crystals of the ice may form leading eventually to rain. For ice $a = 0.454$ nm (4.54 Å), $c = 0.741$ nm (7.41 Å) and the O–O distance is 0.26 to 0.27 nm (2.6 to 2.7 Å); in the iodide, $a = 0.458$ nm, (4.58 Å) and $c = 0.749$ nm (7.49 Å) and Ag–I is 0.281 nm (2.81 Å).

Copper Pyrites-Chalcopyrites

This is a sulphide of copper and iron, $CuFeS_2$, which occurs as tetragonal crystals which are often twinned. Both metal atoms have a coordination number of 4 being surrounded tetrahedrally by sulphide particles just like the zinc atoms in zinc blende. In fact the chalcopyrite structure may be derived theoretically from the blende structure by replacing alternate zinc atoms by copper atoms and then replacing the remaining zinc atoms by iron. Resulting from the relationship between the two structures is the fact that the unit cell of the $CuFeS_2$ is twice that of the blende.

Iron pyrites (FeS₂)

Iron pyrites belongs to the cubic (pyrite-type) system and the crystals are either cubic or have the pentagonal dodecahedral form which receives the name of the pyritohedron as it is the normal form of pyrites. The faces are often striated with the striae of one face at right angles to those of the adjacent faces (Fig. 5.6). The symmetry therefore is not as high as that of

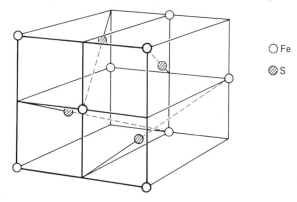

○ Fe

⊘ S

Fig. 5.6. Half of a unit cell of iron pyrites (FeS₂)

rock salt or sodium chloride but possesses trigonal symmetry across the diagonals of the cube. The crystal is not polar along the diagonals as was zinc blende. X-ray diffraction studies show that the lattice dimension is 0·54 nm (5·4 Å) and that there are four molecules of FeS_2 in the cubic cell. X-ray diffraction photographs taken along the (100) and the (110) planes confirm the cubic symmetry. The (111) plane along the diagonals is of particular interest as along this axis is where the trigonal symmetry occurs. In a typical experiment Bragg used rhodium alpha (Rhα) rays for which λ is 0·0614 nm (0·614 Å) and reflections from the (110) plane occurred at values of θ of 3° 14′, 6° 29′, 9° 44′ and 13° 2′.
Where

$$\sin \theta = \frac{n\lambda}{2d}$$

$$\sin \theta = n \times 0·0564$$

Using the simple laws of reflection the ionisation chamber can be set to receive these reflections. The first observation is that the reflections for when n is an odd number are missing. The interpretation for this is that the X-rays from the iron atoms and the sulphur atoms interfere with one another such that for odd values of n they cancel each other out. By noting the positions of the absences from other reflections it is possible to elucidate the structure of iron pyrites completely. Iron pyrites is better known as 'fools gold', because of its golden colour.

116

Quartz

Quartz was known to the ancient Greeks and occurs in a variety of crystalline forms. The quartz form of silica belongs to the hexagonal system and the quartz class. The crystals usually occur as hexagonal prisms which are terminated by positive and negative hexagonal rhombohedra so that the crystals take the form of hexagonal pyramids. The pure crystals are colourless but sometimes there are impurities which give the crystals a colour. Large single crystals of quartz are used for stabilising the wavelengths of radio transmitters. Quartz crystals were one of the original filters used in coaxial circuits for long distance telephoning, each conversation being controlled at a particular frequency by a filter of a single crystal of quartz.

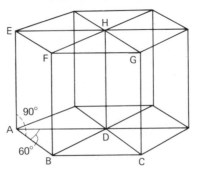

Fig. 5.7. The unit cell of quartz (SiO_2)

More recently crystals such as ethylenediamine tartrate have replaced quartz crystals. Quartz clocks are used as time controls at the Royal Observatory. The clock consists of a quartz crystal which is kept in oscillation by a valve circuit with a frequency of 100 kHz (100 000 cycles per second). The change of time of the clock can be controlled to within one-thousandth of a second in twenty-four hours. As stated, crystals of quartz are hexagonal; the hexagon (Fig. 5.7) is built up from unit cells which have the shape of a parallelepiped ABCDEFGH. The length of AD is 0·488 nm (4·88 Å) and DH or c is 0·537 nm (5·37 Å). The screw axis in quartz is shown by the fact that the reflections from the set of planes parallel to ABCD contain the third orders but not the first and second. The conclusion is that there are effectively three identical equidistant layers within the unit cell itself each separated by one-third of the height of the cell. From preliminary investigations it can be shown that there are three molecules per unit cell and therefore there will be one molecule in each layer, but the overall symmetry of the quartz about the c axis, that is HD, is trigonal. In order that the crystal has this symmetry the three molecules must be arranged in a spiral about the c axis. Each molecule is a silicon dioxide group or SiO_2 group. The trigonal symmetry arises by a combination of a rotation of 120° and a translation of one third of the unit cell (Fig. 5.8). In the first stage, one silicon is moved up one third of the unit cell and rotated through 120°, bringing the second and first units into coincidence. This brings a into coincidence with a_1 and the second operation brings a_1 into coincidence with a_2 and the third brings a_2 into coincidence with a_3 and a_3 is identical to a. Each of the lines Aa, etc., are parallel to one of the faces of the unit cell.

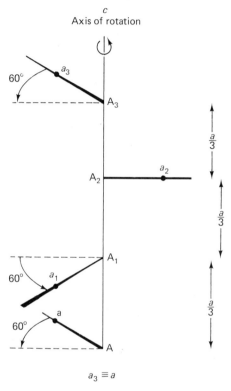

Fig. 5.8. The threefold axis of rotation in quartz

Unfortunately arguments from symmetry alone do not provide sufficient evidence and it is normally necessary now to look at the relative intensity of the spots. At the time of the first investigations of quartz crystals the crystallographers were able to arrive at the true structure by a series of brilliant deductions. Nowadays there are the Fourier equations by which the intensity of the lines may be calculated but the equations are too complicated to be manipulated quickly and a computer is used to work out values of the unknowns in the equation (see Chapers 3 and 7). It turns out that the structure of quartz is related to certain other forms of silica. The quartz crystal exhibits polymorphism, for at room temperature alpha quartz is the stable form, while at higher temperatures beta quartz, which is simpler than alpha quartz, is the more stable form. Tridymite and cristobalite

(Fig. 5.9) are based on arrangements of silicon dioxide units. In all these forms of silica the silicon atom is at the centre of a regular tetrahedron of oxygen atoms but the tetrahedra are usually attached and arranged differently in the different crystal structures. The silicon–oxygen (Si–O) distance is always between 0·152 nm (1·52 Å) and 0·16 nm (1·6 Å) which would emphasise that there is about 50 per cent ionic character although there must be some covalent character in the bond. In all the forms of silica every oxygen atom is shared by two silicon atoms and the result is that the whole crystal is one large molecule or macromolecule. Again it is possible to say you have seen a macromolecule if you have seen a crystal of sand. The structure of quartz is illustrated in Fig. 5.9. Quartz exists in two different forms, one which is dextro- and one which is laevorotatory. Dextro-rotatory means rotating or deviating the plane of vibration of polarised light to the right (observer is looking against the oncoming incident light). Laevorotatory means rotating the light to the left. The difference is that in the two different forms the spirals of silicon dioxide units are different, one is referred to as a righthanded spiral and the other is a lefthanded spiral. The spirals or screw axes are very important and we shall return again to these when biological molecules are considered in Chapter 7.

Silicates

The basis of the structure of the quartz and of the more complex silicates consists of SiO_4 tetrahedra (Fig. 5.10). There are four ways in which the SiO_4 tetrahedra can be linked.

Isolated $[SiO_4]^{4-}$ *tetrahedra.* This occurs in combination with metal ions such as in calcium silicate Ca_2SiO_4. These compounds are found in Portland cement and in lime, in sand mortars and in some plasters.

Tetrahedra linked by sharing one oxygen atom. The result is that an ion $[O_3Si-O-SiO_3]^{6-}$ is formed, this ion is found in several silicates such as the zinc ore haemimorphite $(Zn_4(OH)_2Si_2O_7)$.

Two oxygens shared by the tetrahedra. In this structure the possibility of forming rings or chains arises. Emeralds or beryl contain this type of structure in which six silicon atoms are present in one chain.

When three corners of the SiO_4 *tetrahedra are shared.* Here a sheet structure is quite normal and can extend indefinitely in two dimensions as in the layers of carbon in graphite. These sheets are very common in the structure of micas. Micas are valuable as thermal and electrical insulators and occur as plates which cleave into thinner sheets. Talcum powder is made of talc (French chalk) and owes its lubricating properties to its layer structure, as in graphite the forces between the layers are rather weak, of the order of 20 kJ mol^{-1}, so that the layers are able to slide over one another.

The structure is built up on cubes

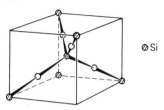

in which there are outer silicon atoms which are joined through oxygen atoms to the central silicon atom. The cubes themselves are arranged tetrahedrally:

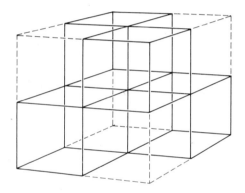

The structure is:

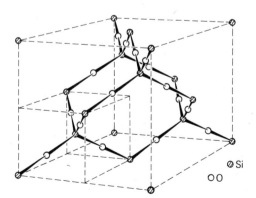

Fig. 5.9. The β cristobalite structure
The structure is based on diamond or zinc blende arrangement of tetrahedra. Can you see how?

Interlinking of Tetrahedra

$[SiO_4]^{4-}$—Silicon surrounded by four oxygens

 Na_4SiO_4—Sodium silicate

$[Si_2O_7]^{6-}$—Two tetrahedra with one linking oxygen

$$\begin{bmatrix} & O & & O & \\ & | & & | & \\ O- & Si & -O- & Si & -O \\ & | & & | & \\ & O & & O & \end{bmatrix}^{6-}$$

$Sc_2Si_2O_7$ Scandium (III) silicate

$[SiO_3]^{2-}$—Infinite chain of tetrahedra—one linking oxygen

$$\begin{array}{ccccccc} & O & & O & & O & & O \\ & | & & | & & | & & | \\ -O- & Si & -O- & Si & -O- & Si & -O- & Si & -O- \\ & | & & | & & | & & | \\ & O & & O & & O & & O \end{array}$$

The pyroxenes
including
$MgSiO_3$; $CaMg(SiO_3)_2$

$[Si_3O_9]^{6-}$—Three tetrahedra with two interlinking oxygens such that there is a ring formed.

$CaSiO_3$ or more correctly $Ca_3Si_3O_9$

$[Si_6O_{18}]^{12-}$

Beryl

Fig. 5.10. The structure of silicates

When the oxygens are all shared by the SiO_4 *tetrahedra,* This is the structure of quartz which we have met before. In one form, that is in the cristobalite, the silicon atoms are arranged in exactly the same way as the carbon atoms are arranged in diamond (Fig. 5.9) but the silicon atoms are linked by oxygens. The silicon atoms, therefore, are placed in a face-centred lattice and in addition there is one silicon atom in the tetrahedral positions within the cube, between each silicon there is an oxygen atom.

Although the framework in which the SiO_4 tetrahedra completely share all the oxygens is electrically neutral it is sometimes possible to replace the silicon atoms by charged metal ions, and such structures are known as framework structures but by replacing Si^{4+} ions by Al^{3+} ions the result is that there is an overall negative charge on the structure. The complete structure could be regarded as one gigantic ion but the charge can be neutralised by inserting an extra positive ion such as sodium or calcium. Examples are:

Zeolite $Na(AlSi_2O_6)H_2O$, Ultramarine $Na_8Si_6O_{24}S_2Al_6$
Felspars $KAlSi_3O_8$ and $Ba(Al_2Si_2O_8)$.

Zeolites contain water in the holes or interstices of the structure and they exhibit a property which is called ion exchange. That is positive ions which are situated along with the water molecules in the interstices may be changed for other positive ions. In this way zeolites may be used for softening water, because the hardness of water is mainly due to calcium and magnesium ions. The zeolite can exchange the sodium ions for the magnesium or calcium ions and in this manner the water is softened. This process is better known as the Permutit method for softening water. The felspars are very important rock forming minerals. Ultramarine was well known during the war as the blue used in laundering clothes and it is possible to alter the colour of the ultramarine by changing the ions for sodium ions or silver ions.

Copper(I) oxide

Copper(I) oxide belongs to the cubic system and the crystals are either octahedral or rhombododecahedral. The cleavage is often poor parallel to

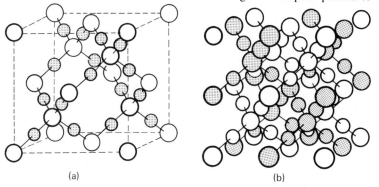

(a)   (b)

Fig. 5.11. Copper(I) oxide

the faces of the octahedron. Rather surprisingly, copper(I) oxide is a good example of a framework structure. Just think a little about the chemistry of copper(I) oxide, it is black and it is very unreactive. There are two series of interpenetrating frameworks which are based on the structure of diamond. Oxygen atoms are placed at the points of the diamond network and a copper atom is placed between each oxygen. A second network, which is illustrated in Fig. 5.11, is also present and is obtained by slightly moving the first lattice. There are no covalent or ionic bonds between the atoms of the two frameworks and they are not able to move over one another as in a typical layer lattice.

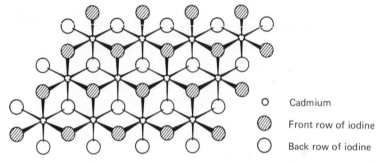

○	Cadmium
◕	Front row of iodine
◯	Back row of iodine

Fig. 5.12. The composite layer of cadmium(II) iodide

Cadmium(II) iodide and Palladium(II) chloride

Cadmium iodide may be referred to as a two dimensional molecule. Earlier in this chapter it was illustrated that diamond was a three dimensional molecule but where a layer lattice is formed it is possible to call this type of crystal structure a two dimensional molecule. Therefore graphite and talc may also be referred to as two dimensional molecules. In the layer structure of cadmium iodide the iodide ions form a close packed structure, the meaning of which will be discussed later. The cadmium ions fit into the structure such that each cadmium ion is surrounded by six iodide ions and part of a layer is illustrated in Fig. 5.12. The metal atoms which are in the plane of the paper occupy the holes or interstices between the iodine atoms, which in turn lie in two parallel planes above and below the plane of the cadmium ions. The effect is, therefore, that there is a plane of iodide ions and a plane of cadmium ions. The ratio of the number of cadmium ions to the number of iodide ions is 1:2 but if one third of the metal atoms are removed from this kind of layer lattice the ratio of the number of ions is 1:3. This second type of layer lattice is the basis of the structures of the trihalides chromium (III) chloride and bismuth(III) iodide, chromium(III) bromide and iron (III) chloride ($FeCl_3$). They are all characterised by having close packed arrangement of the halogen ions or atoms.

There is effective contact between the atoms in a metal for two reasons. First, the atoms are considered to be ionised and the 'electron cloud' between the atoms constitutes the metallic bond and holds the 'atoms' together. The negative charges on the oxygen ions repel one another and there is no effective contact between the negative ions. Therefore the average distance between the ions is greater in the oxides than in the metals. Secondly, the ionic radii are decreased when an electron is removed to form a negative ion but increased when an electron is added to form a negative ion. Some values are Na^+ 0·095 mn (0·95 Å); Mg^{2+} 0·065 mn (0·65 Å); Al^{3+} 0·05 nm (0·50 Å) and O^{2-} 0·14 nm (1·40 Å). The nucleus can hold electrons especially when there are less electrons than protons in positive ions but electrons are loosely held when there are more electrons than protons, as in negative ions. Oxides are more open than metals.

Palladium(II) chloride is of interest to inorganic chemists because the palladium is surrounded not only by four chloride units in the same plane but also by two other units at a much greater distance above and below the plane. The reason is that the palladium atoms are directly bonded to the four chlorines but the layers stack so that the chlorides are not above the other similar units. This compound exhibits what we call a tetragonal distortion to give four short and two long bonds from the regular cubic structure in which all six bonds would be identical.

In sapphire or corundum, that is alpha alumina (αAl_2O_3), there is a close packed arrangement of oxygen atoms or ions and the aluminium atoms fit in between the oxygens. In fact the aluminium ions are in two thirds of the holes in which they can be surrounded by six oxygen atoms and the formula for aluminium oxide is Al_2O_3. Jewellers rouge or haematite which is αFe_2O_3 (or alpha ferric oxide) is of a similar structure to sapphire. Certain mixed crystals, that is of $KF.MgF_2$ and $CaO.TiO_2$ are built up also of close packed networks of fluoride and oxygen atoms respectively, the metal atoms fitting in the holes left by the atoms. Many minerals, such as the mineral $CaTiO_3$ are also built up of a close packed network of oxygen atoms.

Halogens

These form dimeric molecules which are still present in the structure of the molecular crystal. In the molecular structure of iodine the molecules fit together in layers and the layers fit over one another to give a three dimensional structure. The covalent distance between the nuclei is 0·267 nm (2·67 Å) so that the atomic radius of iodine is 0·13 nm (1·34 Å) but the distance between the layers is much larger 0·43 nm (4·3 Å). It might be thought that this latter distance also would be the intermolecular distance but this distance is only 0·406 nm (4·06 Å). But why is this distance smaller than would be expected? The reason is that there is some degree of covalent bonding between the molecules within a layer. The flakiness of iodine arises because the interlayer forces are so weak. Iodine may be sublimed at low temperatures because in this process in which individual molecules are formed the weak inter- and intralayer forces are easily overcome. The intralayer forces in iodine are stronger than in crystals of chlorine or bromine which may be formed at low temperatures but all three have similar structures. There is naturally a correlation between the bond and the normal chemistry of iodine because iodine is associated in benzene, has metallic lustre (possibly indicating that the electrons are partially delocalised) and iodine has much higher stable valency states (IF_7) than chlorine or bromine.

Group 6B

Oxygen is normally written $O=O$ but even this simple piece of chemistry must go because we now recognise that oxygen is paramagnetic having two unpaired electrons. The gas liquid and solid phases all contain dioxygen (O_2) but polymorphism occurs in the solid phases because there are several different crystalline forms at different temperatures owing to the different distributions of the dioxygen molecules. Sulphur, as most of us know, exhibits allotropy in which the orthorhombic form which is stable at room temperature consists of S_8 puckered rings. The unit cell is very complex as it involves the packing of S_8 units which usually align themselves into straight columns resulting in the formation of layers of S_8 molecules. When this solid is just melted it is the intermolecular forces which are overcome so that the liquid contains S_8 molecules. But when this liquid is cooled it is the monoclinic sulphur which is formed which also consists of S_8 rings but packed in a different manner from orthorhombic sulphur. On standing the molecules realign themselves to give rhombic sulphur without altering the external symmetry and here is a case where the external structure does not give a clue to the internal structure. When very hot molten sulphur is suddenly cooled the product is plastic sulphur which consists of long chains of sulphur atoms and the chains are at all conceivable orientations to one another. The internuclear covalent distance is 0·212 nm (2·12 Å) which is the covalent diameter of sulphur. Although the structures of the solid forms are quite well characterised the structure of the liquid is still being investigated. When orthorhombic sulphur is slowly heated the rings remain intact when the solid melts but on further heating the rings break open giving long sulphur chains which make the sulphur go red and thicken. These chains break down at higher temperatures giving S_6 rings as in rhombohedral sulphur.

Selenium, like sulphur, also forms Se_8 rings and infinite chains but the chains are far more stable than the rings so that the crystalline structure of selenium consists of infinite chains of selenium atoms. The result is that there is a helical chain giving a 'one dimensional' molecular structure. Tellurium is similar to the metalloid selenium but αpolonium is considered to have the simple cubic structure. Normally one says that the metallic character of the elements increases down a group but how can this be reconciled with the structures? In a metallic structure there is a three dimensional arrangement in which the coordination number of the metal atom is often as high as 12 but in three elements the coordination number is 2. The answer is that although sulphur, selenium and tellurium are only

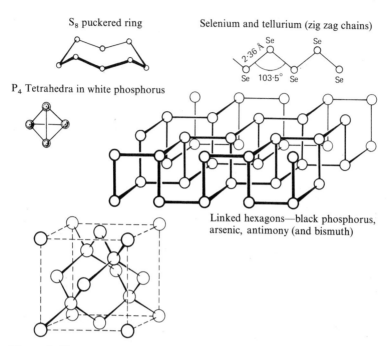

S₈ puckered ring

Selenium and tellurium (zig zag chains)

P₄ Tetrahedra in white phosphorus

Linked hexagons—black phosphorus, arsenic, antimony (and bismuth)

Diamond, silicon, germanium and grey tin

Fig. 5.13. Structures of some elements of Group 4b, 5b, 6b

attached to two atoms the remaining atoms in the chains are nearer to tellurium than to sulphur. For example the covalent diameters are S 0·212, Se 0·236 and Te 0·282 nm while the van der Waal diameters, which are the distances to the next chains are 0·33, 0·346 and 0·346 nm respectively. Polonium is dimorphic, the alpha form is possibly simple cubic and the beta form belongs to the trigonal system.

Group 5B

The elements of Group 5 consist of nitrogen, phosphorus, arsenic and antimony and bismuth. Of these nitrogen is a gas at room temperature. Like sulphur and carbon, phosphorus exhibits allotropy. The three main allotropes are white, red and black phosphorus. Each of these is polymorphic, and altogether eleven modifications are known. In the liquid phase phosphorus and solid white phosphorus exists as P₄ tetrahedra, but commercial red phosphorus has an uncertain crystalline structure. Black phosphorus which is obtained in the crystalline form by heating white phosphorus under high pressure in the presence of mercury as a catalyst has a series of corrugated sheets of phosphorus atoms. Each phosphorus is bound to three neighbours and the structure is made up of layers of hexagons. The crystals are flaky, like graphite and the forces between the layers are the weak van der Waal bonds. The normal forms of arsenic, antimony and bismuth are bright and metallic and have crystal structures similar to that of black phosphorus. Unstable allotropes of arsenic and antimony contain the tetrahedra of the respective atoms.

Arsenic and antimony resemble phosphorus in that they exhibit polymorphism but the metallic form becomes more stable as the atomic weight increases. Bismuth does not have a stable crystalline form which contains Bi_4 tetrahedra. The stable crystalline forms of grey arsenic, grey antimony and bismuth are regarded as semi-metallic and in these structures there are layers of macromolecules which in one way resemble graphite as there are hexagons of the element. Unlike graphite, however, there are no free electrons and the interlayer distance decreases from phosphorus to bismuth. The covalent diameters are P 0·218; As 0·25; Sb 0·29; Bi 0·31 nm but the distance to the nearest neighbouring atom in the next layer is P 0·368; As 0·315; Sb 0·336 and Bi 0·347 all values in nm.

Group 4B

We have already discussed the structures of diamond and graphite but although the stable structures of silicon, germanium and grey tin have the diamond structure there are no comparable forms to graphite, as none of these elements is found with coordination number 3. Metallic white tin has a complex structure in which each tin atom is bonded to six others, and owing to the inert pair effect it is probable that only two electrons are used in the conduction band. With lead there is a completion of the trend to metallic properties as the structure is cubic close packed.

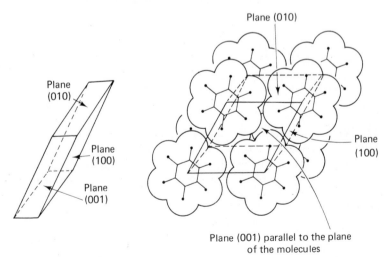

(a) the shape of the crystal (note the (001) (100) and (010) faces)

(b) the positions of the molecules related to the crystal

Fig. 5.14. The structure of hexamethylbenzene $C_6(CH_3)_6$

Hexamethyl benzene

For over a century the structure of benzene had been a problem to chemists. On the basis of certain evidence it was thought most probable that the benzene structure was based on a ring of six carbon atoms. It was not until the structure of hexamethylbenzene was studied in detail that there was any confirmation from X-ray diffraction data regarding the structure of benzene. The reason why benzene itself was not studied initially was that X-rays are diffracted by the electrons around the atom and therefore it is difficult to locate the positions of hydrogen atoms in the presence of carbon atoms. Several questions must be asked regarding the structure of benzene. For example, is the organic molecule a separate entity in the crystalline state? If the carbon atoms in benzene are arranged in the form of a ring, is the ring hexagonal and what is the size of ring? Is the ring planar or is it puckered? The first problem was to grow crystals of hexamethyl benzene, the shape of which is shown in Fig. 5.14. Crystallographers take Laue photographs with the incident beam along a known axis (or other directions for particular purposes) or rotation photographs using particular radiation. The crystal rotates about a crystallographic axis with the axis perpendicular to the incident beam and parallel to the X-ray film. The rotation photographs enable one to calculate the structure projections along the axis in question. By refining the technique in the manner previously illustrated the positions of the molecules could be related to the crystal structure. The structure confirmed that the ring of benzene was not only hexagonal and planar but that the carbon-carbon distance was intermediate between that expected for a single and a double bond.

Plate 5.2. Molecular sieves

Molecular sieves

These are commonly used as drying agents for solvents and can act in one or both of two ways. Firstly reference to Plate 5.2 will show that there are tubes, the walls of which are made or formed by the molecules of the sieve

128

and these tubes allow the small water molecules to move in where they are trapped. The larger solvent molecules such as benzene are unable to move into the tubes. Secondly the sieves provide polar charged surfaces which attract the polar water molecules and do not attract the non-polar solvent molecules.

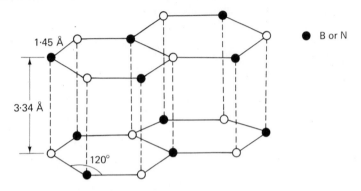

Fig. 5.15. Special boron nitride (BN)—graphitic
Commercial boron nitride does not usually have this structure but there is a special boron nitride which is graphitic

So far then we have looked at quite a few crystals and it is now time to see if these structures fall into any type of pattern. One of the ways in which Goldschmidt classified compounds was according to the type AX, for example caesium chloride and sodium chloride are typical compounds of the type AX. It is useful to refer to these according to their coordination number (Fig. 5.16). Coordination number 1 is typical of single molecules such as oxygen and chlorine. Coordination number 2 is typical of molecular chains, e.g. the chains of sulphur. Coordination number 3 is typical of a boron-nitride structure (Fig. 5.15). Coordination number 4 is typical of the zinc blende lattice and coordination number 6 is typical of sodium chloride and of nickel arsenide which will be dealt with in Chapter 6. Coordination number 8 is typical of caesium chloride. Compounds of AX_2 are listed in Fig. 5.17 and in addition one or two new structures which we have not dealt with are considered and the reader is referred to more advanced texts to study their structure.

One question which we must ask, however, is 'To what extent are ionic and crystal radii constant?' (Fig. 5.18). On the whole, when there is a change in lattice type there is only a small percentage change in the ionic radius. There is always a decrease of ionic radius when the coordination number decreases, the decreases being small compared with the inter-ionic and inter-molecular distances. Larger differences occur when the ions or atoms have not the inert gas structure. The departure of ionic compounds from the idea of incompressible spheres is represented in the theory of polarisation.

Coordination number	Types
1	Single molecules (Cl_2, N_2, CO). Molecular structures (solid N_2)
2	Sulphur in chains of sulphur. Structures of many chains
3	Structures of the boron nitride type
4	Structures of the zinc blende or wurtzite (ZnS) type and tetragonal layer structures
6	Structures of the sodium chloride and nickel arsenide types
8	Structures of the caesium chloride type

Fig. 5.16. Coordination type in AX crystals

Numbers of coordination	Types
2 and 1	Single molecules and molecular structures
4 and 2	Structures of α and β quartz, α tridymite, α cristobalite and cuprite types
6 and 3	Structures of anatase, rutile, cadmium iodide, molybdenite types
8 and 4	Structures of fluorite type

Fig. 5.17. Coordination type in AX_2 crystals

Structure transition	Change in the coordination number	Change in the inter-ionic distances
Caesium chloride type to sodium chloride type	8 to 6	Decrease of 3 per cent
Sodium chloride type to zinc sulphide type	6 to 4	Decrease of 5 to 8 per cent
Fluorite type to rutile	8 and 4 to 6 and 3	Decrease of 3 per cent

Fig. 5.18. Inter-ionic distances decrease with coordination number

Fajan's Rules

Fajan's Rules illustrate the tendency of certain compounds to form ionic or covalent bonds. In general the ionic radius is different from the covalent radius. An ionic compound is formed when the cation, that is the M^{n+} species, has high volume and low charge, and the anion should have low volume and a small charge. Conversely a covalent compound is formed when the cation has small volume and high charge and the anion has high charge and large volume (Fig. 5.19). The reason for these changes is connected with the polarising power of M^{n+} or cationic species and the polarisability of anions. The polarising power is defined as the ability of positively

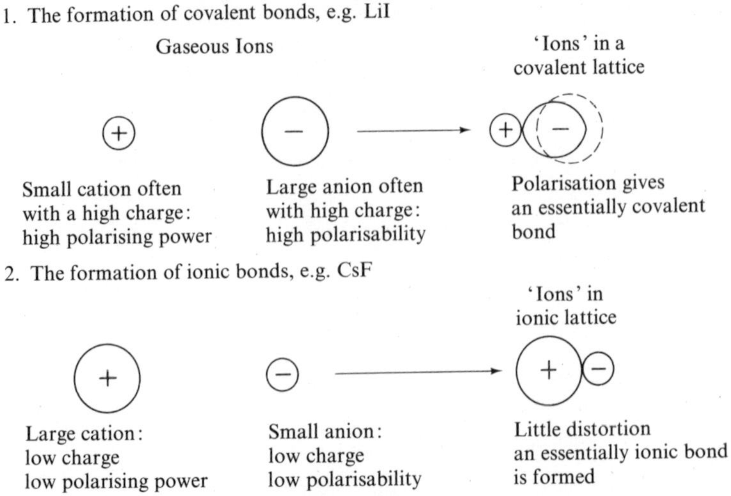

1. The formation of covalent bonds, e.g. LiI

Gaseous Ions

'Ions' in a covalent lattice

Small cation often with a high charge: high polarising power

Large anion often with high charge: high polarisability

Polarisation gives an essentially covalent bond

2. The formation of ionic bonds, e.g. CsF

'Ions' in ionic lattice

Large cation: low charge low polarising power

Small anion: low charge low polarisability

Little distortion an essentially ionic bond is formed

Fig. 5.19. An illustration of Fajan's Rules

charged ions to draw the negative electrons from anions. Now in ionic bonds there is complete transfer of electrons from the cation to the anion but if the cation is able to draw this negative charge from the anion then the electron is not associated completely with the anion. A covalent compound is formed when the cation is able to polarise the anion. The polarisabilities (Fig. 5.20) represent the ease with which the electrons may be withdrawn from the anions. For example, the polarisability of the iodide ion is much greater than that of the fluoride ion, which means that the electrons may be withdrawn from the large iodide ion much more easily than they can from the small fluoride ion. The reason is that the electrons are very closely associated with the positive nucleus in the fluoride ion but less attracted to the nucleus in the iodide ion. Iodides tend to form covalent compounds whereas fluorides and oxides tend to form ionic compounds.

The covalent radii of the elements (Fig. 5.21) vary with atomic number and increase down any given group and decrease across a given period, for example the covalent radius of lithium 0·12 nm (1·2 Å) is greater than that

131

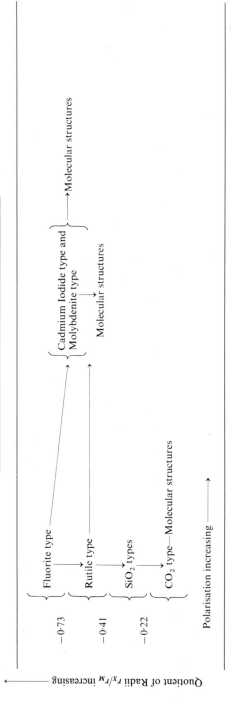

Fig. 5.20. Polarisation and radius ratio

Fig. 5.21. Covalent radii of the elements (Å)

	Ia	IIa	IIIa	IVa	Va	VIa	VIIa	VIII			Ib	IIb	IIIb	IVb	Vb	VIb	VIIb	0
	1 H 0·4																1 H 0·4	2 He 0·5
	3 Li 1·5	4 Be 1·1											5 B 0·8	6 C 0·8	7 N 0·5	8 O 0·6	9 F 0·7	10 Ne
	11 Na 1·9	12 Mg 1·6											13 Al 1·4	14 Si 1·2	15 P 1·1	16 S 0·9	17 Cl 0·9	18 A
	19 K 2·3	20 Ca 1·8	21 Sc 1·6	22 Ti 1·4	23 V 1·3	24 Cr 1·2	25 Mn 1·4	26 Fe 1·2 / 27 Co 1·2 / 28 Ni 1·2			29 Cu 1·3	30 Zn 1·3	31 Ga 1·2	32 Ge 1·2	33 As 1·2	34 Se 1·2	35 Br 1·1	36 Kr
	37 Rb 2·5	38 Sr 2·2	39 Y 1·8	40 Zr 1·6	41 Nb 1·4	42 Mo 1·4	43 Te 1·3	44 Ru 1·3 / 45 Rh 1·3 / 46 Pd 1·4			47 Ag 1·4	48 Cd 1·5	49 In 1·6	50 Sn 1·5	51 Sb 1·4	52 Te 1·4	53 I 1·3	54 Xe
	55 Cs 2·7	56 Ba 2·1	57 La 1·9	72 Hf 1·7	73 Ta 1·4	74 W 1·4	75 Re 1·4	76 Os 1·3 / 77 Ir 1·3 / 78 Pt 1·4			79 Au 1·4	80 Hg 1·5	81 Tl 1·7	82 Pb 1·7	83 Bi 1·5	84 Po 1·7	85 At	86 Rn
	87 Fr 2·8	88 Ra	39 Ac 1·9															

58 Ce 1·8	59 Pr 1·8	60 Nd 1·8	61 Pm 1·8	62 Sm 1·8	63 Eu 1·9	64 Gd 1·8	65 Tb 1·8	66 Dy 1·7	67 Ho 1·7	68 Er 1·7	69 Tm 1·7	70 Yb 1·9	71 Lu 1·7
90 Th 1·8	91 Pa 1·6	92 U 1·4	93 Np 1·3	94 Pu 1·5	95 Am	96 Cm	97 Bk	98 Cf					

10 Å = 1 nm

of fluorine 0·072 nm (0·72 Å) the reason being that the electrons are closely associated with the nucleus in fluorine as there are more protons in the nucleus of fluorine than in lithium. The covalent radius of lithium may be measured by electron diffraction not X-ray diffraction as Li_2 molecules are present to a certain extent in the gaseous phase, but not in the solid phase. We have seen how Goldschmidt assigned a value to the ionic radius of an element and a similar procedure may be used for atomic or covalent radii. The normal standard is the C–C distance in diamond 0·154 nm (1·54 Å) which is closely related to the dimensions of the unit cell. In methane the C–H distance may be found and subtraction of the radius of the carbon atom gives the covalent radius of hydrogen 0·114 (1·14) minus 0·077 (0·77) gives 0·037 nm (0·37 Å). The values of the covalent radii are usually taken to be half the distance between the identical atoms if possible in a homopolar diatomic molecule and are often quoted as atomic radii as the size of individual atoms is very difficult to determine. In general atomic radii are larger than the radii of the same atom with a positive charge because the charge means that the negative electrons are held more tightly than they are in the neutral atom. Currently there are two values which are quoted by books for the atomic radius of gold namely 0·144 and 0·137 nm but this second value is probably wrong because the ionic radius of the Au^+ ion is 0·136 nm.

Questions

1 Draw the structure of (a) iron(II) pyrites (FeS_2); (b) copper(I) oxide; (c) cadmium(II) iodide; (d) graphite; (e) alpha quartz.

2 Write down the coordination types for AX and AX_2 compounds.

3 Draw the structures of as many precious stones as you can.

4 What is the coordination number of carbon in (a) calcite; (b) graphite; (c) diamond?

5 Compare the ionic radii in Chapter 4 with covalent radii in Fig. 5.21.

6 Draw clearly, with examples, crystal lattices of (a) layer lattices; (b) framework lattices.

7 The structure of silicates is related to tetrahedra of $[SiO_4]^{4-}$. Draw, with examples, the structure of as many silicates as you can.

8 Indicate the way in which X-ray diffraction studies elucidate the structures of (a) zinc blende; (b) iron pyrites; (c) alpha quartz; (d) hexamethylbenzene.

9 To what extent does crystallography explain the chemistry of (a) copper(I) oxide; (b) diamond and graphite; (c) silicates?

10 The fluorides of barium, strontium and calcium have the fluorite structure whereas magnesium fluoride has the rutile structure. Account for this difference in structure.

11 (a) Distinguish between the words lattice and structure; (b) Discuss the screw axis in quartz. (c) Why is calcite birefringent? (d) Why do spots have different intensities? (e) What are the structures of N_2, C_6H_6, PCl_5, $CsNO_3$, $LiNO_3$, LiI, PBr_5, ZnS?

12 Why are the following statements incorrect? (a) The unit cell of diamond consists of one carbon atom surrounded tetrahedrally by four carbon atoms? (b) The sodium chloride lattice is body centred cubic? (c) Caesium fluoride has a body centred structure. (d) Zeolites are interstitial compounds. (e) Plastic sulphur contains rings of sulphur. (f) Boron nitride has the same structure as diamond. (g) Palladium(II) chloride has a molecular structure. (h) Naphthalene has a layer structure. (i) Caesium chloride exhibits 6:6 coordination. (j) Zinc blende and diamond belong to the same space group. (k) In graphite carbon has a coordination number of six. (l) X-rays have a wavelength of 0·154 nm. (m) Calcium chloride has a close-packed structure. (n) Nickel arsenide, which exhibits 6:6 coordination, is isomorphous with sodium chloride the molecules of which are transparent.

6 The metallic state, alloys, piezo-electric effect, field-ion microscope, masers and lasers

The study of metals provides metallurgists and chemists with many interesting problems. For example, 'What is the structure of metals? Does the structure explain why metals are very good conductors of electricity and heat? Can the structure explain why conductivity decreases rapidly as impurities are added? Why are metals very good catalysts for many reactions? Why do X-ray diffraction spectra become diffused when the metal is stressed? Any proposed structure of metals must be able to answer these questions and to account for their malleability and ductility together with the fact that different ones are able to alloy with each other. The metallic bond must have the following properties:

1. The bond must be able to act between identical atoms and sometimes between atoms of differing types. Experimentally it has been found that the only limit is that the majority of atoms in the alloy must be metallic; otherwise the alloy loses its essentially metallic properties.

2. The bond must be undirected because the properties of metals are usually approximately the same. X-ray diffraction shows that the coordination number of metal atoms is very high and is normally 12 or 8.

3. The strength of the metallic bond must vary inversely as some high power of the internuclear distance. The alkali metals which have high atomic volumes have low melting points and platinum which has a low atomic volume has a very high melting point and boiling point.

4. The bond must allow the transfer of electrons from one atom to another to account for the conductivity of electricity and heat.

The boundary between metallic, covalent and ionic structures (Fig. 6.1) is not always sharp but at the extremes of each case one can recognise that sodium chloride is ionic, hydrogen is covalent and copper is a metal. In between these three extremes, however there are different compounds or alloys in which the bonding is intermediate between the respective types. Hydrogen fluoride contains the hydrogen-fluorine bond which is partially covalent and partially ionic. The alloy Cu_3Sn has bonding which is intermediate between metallic and covalent.

The atoms in pure metals are all identical, and the structure of the metal may be considered to be formed of layers of these atoms. In the first layer each atom is surrounded by six others (Fig. 6.2) and this layer forms the basis from which further layers may be formed (just as we saw earlier when a crystal was grown). The second layer settles over the holes in the first layer, there are two equivalent ways in which the second layer may form over the first layer. The third layer has two choices, however, that is whether or not to go over the atoms in the first layer. If the third layer is over the atoms in the first layer, then the structure is hexagonal with ABABAB repetition of the layers. If the atoms in the third layer are not over the atoms of the first layer then the repetition is ABCABC or face-centred cubic. The atoms in the fourth layer being directly over those in the first layer. Most metals have these two types of structure which are referred to as '*close packed structures.*' The coordination number is 12 for both cubic and hexagonal close packing. In addition some metals, e.g. tungsten, have a body-centred structure in

136

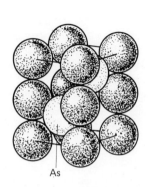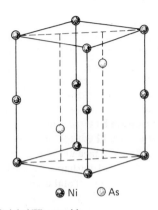

As   ● Ni   ◎ As

Fig. 6.1(*a*). The structure of nickel(II) arsenide

The hexagonal unit cell contains nickel and arsenic atoms each having coordination number six. Each nickel is surrounded octahedrally by six arsenic atoms and each arsenic is surrounded by a trigonal prismatic arrangement of six nickel atoms. The bonding is essentially metallic as there is a wide range of Ni : As ratios. The structure of 'CrS' and many defect lattices are also based on the nickel arsenide structure

Ionic (Sodium Chloride) and Metallic (Nickel(II) Arsenide) Compounds

Radius of Divalent cation in Å	Ca	Mn	Fe	Co	Ni
	1·06	0·91	0·83	0·82	0·78
O	NaCl	NaCl	NaCl	NaCl	NaCl
S	NaCl	NaCl	NiAs	NiAs	NiAs
Se	NaCl	NaCl	NiAs	NiAs	NiAs
Te	NaCl	NiAs	NiAs	NiAs	NiAs
As	—	NiAs	NiAs	NiAs	NiAs
Sb	—	NiAs	NiAs	NiAs	NiAs

Fig. 6.1(*b*). Ionic (sodium chloride) and metallic (nickel(II) arsenide) compounds

Fig. 6.2. The manner in which close packed layers build up, for example in metals

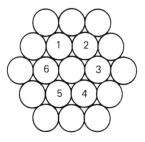

1. The first layer. Each atom is surrounded by six others

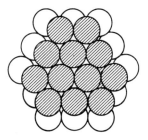

2. The second layer settles over the spaces in the first—there are two equivalent sets of holes

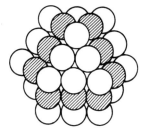

Over the spaces

Over the atoms

3. The third layer has two choices: whether to be over the atoms in the first layer or not.

In order to visualise the alternatives three or more polythene sheets may be taken on which one may draw touching circles on each sheet of polythene. Place one sheet firmly and then put the other sheet in the correct position. The third layer may go into either of the correct positions.

which the coordination number is 8 and this structure is not so close packed. The fact that many metals have close packed structures is indicative that many metal atoms behave as if they were spheres. The reader can make for himself the structures of some metals by using marbles or table tennis balls as representative of the metal atoms. The best way to carry out the construction will be to make the first layer and then carefully add the second layer and then the third layer in the two possible ways. Sometimes it is possible to stick the balls together so that the structure may be kept for reference. Metals such as copper, silver, gold and gamma iron (γFe) have the face-centred cubic structure (Fig. 6.3). Metals such as beryllium, magnesium, alpha zirconium, eta copper and zinc have the hexagonal close packed structure. Sodium, potassium, molybdenum, tungsten and alpha iron (αFe) have the body-centred cubic structure and alpha polonium (αPo) possibly has the single cubic structure. In earlier chapters we have seen that crystal faces contain the maximum number of atoms, and are related to certain arrangements of the atoms within the unit cell (Fig. 6.3). Indeed the planes of the crystal face may be found by finding the planes containing the maximum number of atoms, thus there is a face of the

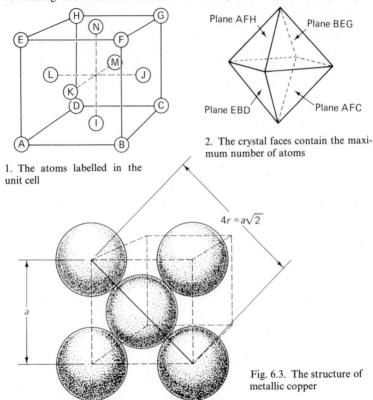

1. The atoms labelled in the unit cell

2. The crystal faces contain the maximum number of atoms

Plane AFH   Plane BEG

Plane EBD   Plane AFC

$4r = a\sqrt{2}$

Fig. 6.3. The structure of metallic copper

crystal of copper which is parallel to planes BEG, etc. The crystal of copper is octahedral, which we have seen is based on cubic symmetry. The close packed hexagonal structure of zinc (Fig. 6.4) is slightly more open than the normal close packed lattice. The arrangements of the atoms are exactly the same as would be expected but the lattice is more open in the respect that the atoms are further apart. The hexagonal structure indicates how the arrangement of the atoms in the first layer is related to the atoms in the third layer. In the first or lower layer one atom is drawn surrounded by six atoms, then three atoms are placed above the holes between these atoms, the third layer is placed directly over the first layer and this gives the hexagonal symmetry, each atom is attached to twelve others, six in the same plane, three above and three below.

The structure of metals generally provides an explanation as to the malleability and ductility of the metal. When a stress is applied to a metal, the metal does not immediately cleave or fracture but the planes are able to move over one another, that is metals deform along the glide planes. When the X-ray diffraction pattern becomes diffuse as the metals are stressed it is because the atoms are distorted from ideal positions. In the case of zinc, after a stress has been applied, the structure would no longer be ABABAB, but would be distorted and of lower symmetry. A distorted arrangement of the atoms leads to a more diffuse X-ray diffraction spectrum. The malleability and ductility of the metals is explained by the fact that the atoms are easily able to move over one another in the glide planes.

Calculations of densities

For metals there exists a simple relationship between the unit cell and the internuclear distance (d) between the particles and the value (Z) of the number of atoms per unit cell. For any cubic cell

a particle at the corners is shared by 8 unit cells,
an edge is shared by 4 unit cells,
a face is shared by 2 unit cells,
the middle of the cell is unshared.

The following arguments are for metals and in which all the atoms are identical and may be regarded as spheres.

Data books usually quote metallic radii which are used in calculating the density of metals. For the face-centred structure (Fig. 6.3):

$$a\sqrt{2} = 4r \qquad a = \frac{4}{\sqrt{2}}r = 2\sqrt{2}r$$

Therefore

$$V_{uc} = a^3 = [2\sqrt{2}r]^3 = 8 \times 2\sqrt{2}r^3 = 16\sqrt{2}r^3$$

There are 8 particles at the corners shared by 8 unit cells. There are 6 particles in faces shared by 2 unit cells. The total therefore is $1 + 3 = 4$.

$$\text{Density} = 1{\cdot}6604\frac{ZM}{V_{uc}} = \frac{1{\cdot}6604 \times 4M}{16\sqrt{2}\,r^3} = 0{\cdot}2935\frac{M}{r^3}$$

In practice this density is also the density of the hexagonal close packed structure but the volume of the unit cell and the number of particles per unit cell are different. In the body-centred structure the density may be calculated but this is not a close packed structure (Fig. 6.4b)

$$4r = a\sqrt{3}$$

$$V_{uc} = a^3 = \left(\frac{4r}{\sqrt{3}}\right)^3 = \frac{64}{3\sqrt{3}}r^3$$

There are two atoms per unit cell as there are eight at the corners and one in the centre.

$$\text{Density} = 1{\cdot}6604\,\frac{ZM}{V_{uc}} = \frac{1{\cdot}6604 \times 2 \times 3\sqrt{3}\,M}{64r^3} = 0{\cdot}2696\,\frac{M}{r^3}\ \text{g/cm}^3$$

$$\text{For }\alpha\text{ iron} = \frac{0{\cdot}2696 \times 55{\cdot}9}{1{\cdot}91} = 7{\cdot}96\ \text{g/cm}^3$$

The experimental density is $7{\cdot}87$ g/cm^3 so that the experimental density is less than the calculated value because there are defects as we have seen in the real structure. For sodium:

$$\text{Density} = 0{\cdot}2696\,\frac{23}{6{\cdot}43} = 0{\cdot}97\ \dot{\text{g}}/\text{cm}^3$$

which is the same as the experimental value showing that there are very few defects in the structure of sodium.

In alloys the approach has to be altered slightly because the particles are of different size and the treatment may be similar to that for MX compounds, e.g. caesium chloride. Let r_1 and r_2 be the Goldschmidt radii of the two ions. In order to calculate Z we must realise that we are finding the number of empirical formulae (CsCl) per unit cell. For caesium chloride (Fig. 6.4d)
There are 8 corner Cl atoms shared by 8 unit cell: $= \frac{8}{8} = 1$
There is one unshared Cs atom: $\qquad\qquad\qquad = 1$
  Total number of CsCl units $\qquad\qquad\qquad = 1$
and there must be an equal number of Cl and Cs atoms for electrical neutrality. The number of CsCl units per unit cell is 1.

$$\text{Density} = 1{\cdot}6604\,\frac{ZM}{V_{uc}} = \frac{1{\cdot}6604.1.M3\sqrt{3}}{8(r_1 + r_2)^3} = \frac{1{\cdot}078M}{(r_1 + r_2)^3}\ \text{g/cm}^3$$

Thus with CsCl $M = 168{\cdot}4$

$$\text{Density} = \frac{1{\cdot}078 \times 168{\cdot}4}{(1{\cdot}67 + 1{\cdot}81)^3} = \frac{1{\cdot}078 \times 168{\cdot}4}{42{\cdot}1} = 4{\cdot}31\ \text{g/cm}^3$$

But the true experimental density is $3{\cdot}97$ g/cm^3 showing that there are some defects in the structure.

In CaF$_2$ the cell diagonal has a length $2 \times 2(r_1 + r_2)$

Therefore
$$a\sqrt{3} = 4(r_1 + r_2)$$
$$a = \frac{4}{\sqrt{3}}(r_1 + r_2)$$
$$V_{uc} = a^3 = \frac{64}{3\sqrt{3}}(r_1 + r_2)^3$$

There are 8 corner fluoride ions, therefore average   = 1
There are 12 edge fluoride ions, therefore average   = 3
There are 6 face fluoride ions, therefore average   = 3
There is 1 unshared fluoride ion, therefore average = 1
Total number of F^- ions per unit cell = 8

There are 4 unshared Ca^{2+} ions per unit cell, therefore $n = 4$ as there are 4 CaF_2 per unit cell.

Again,

$$\text{Density} = \frac{1 \cdot 6604 \, ZM}{V_{uc}} = \frac{1 \cdot 6604 \times 4 \times M \times 3\sqrt{3}}{64(r_1 + r_2)^3}$$

$$= \frac{0 \cdot 5392M}{(r_1 + r_2)^3} \text{ g/cm}^3$$

For CaF_2, $M = 78 \cdot 1$

Therefore

$$\text{Density} = \frac{0 \cdot 5392 \times 78 \cdot 1}{(0 \cdot 99 + 1 \cdot 36)^3} = 3 \cdot 24 \text{ g/cm}^3$$

But the experimental value is $3 \cdot 18$ g/cm³ which gives us a clue that the defects we have talked about are quite common.

One question we might ask is, 'How much free space is there in metals?' The volume of a spherical atom is $2 \cdot 88 \times r^3$. From measurements of atomic volumes, for both the face-centred and the hexagonal close packed structures the average volume per metal atom of radius r is $3 \cdot 04 \, r^3$. So that about 25 per cent of the structure is free space. Ideally the axial ratio of $c:a$ for the hexagonal cell is $1 \cdot 6333$ but we have already seen that for zinc there is a more open structure and consequently there will be rather more than 25 per cent free space. Often departures from the true axial ratio means that the atoms should be regarded as ellipsoids rather than spheres. In the body centred cubic structure the average volume per spherical metal atom is $6 \cdot 16 \, r^3$, which means that on average there is 32 per cent free space.

Calculations of the percentage free space

In fact it is quite possible to calculate the percentage space that is occupied by the atoms and a knowledge of the free space is important in determining the nature of interstitial compounds and interpenetrating structures. In the face-centred cubic structure; (Fig. 6.3c)

Volume of the unit cell $= (1{\cdot}41d)^3$ cm^3

volume unoccupied by one atom $= 4/3\pi(d/2)^3$ cm^3

volume occupied by four atoms in one unit cell $= 16/3\pi(d/2)^3$ cm^3

percentage space occupied by atoms $= \dfrac{16 \times 3{\cdot}14 \times d^3 \times 100}{3 \times 2^3 \times (1{\cdot}41)^3 \times d^3}$

$$= 74 \text{ per cent}$$

Note that the length of the cell is $1{\cdot}41d$ where d is the diameter of the atom regarded as hard spheres and there is 26 per cent free space.

In the body-centred cubic structure there is one full atom at the centre and two half atoms along the diagonal which has a length $2d$ and the dimensions of the unit cell are found using Pythagoras's theorem:

$$(2d)^2 = (\sqrt{2}a)^2 \times a^2$$

$$a = \frac{2}{\sqrt{3}}d$$

$$V_{uc} = \left(\frac{2d}{\sqrt{3}}\right)^3 = \frac{8}{3\sqrt{3}}d^3$$

The number of atoms per unit cell is found by noting that there are eight atoms at the corners which are shared between 8 cells giving one atom per unit cell. In addition there is one at the centre of each cell giving an average of 2 per unit cell.

The number of atoms $= 8$ at corners shared by $8 + 1$ at centre

$$= \frac{8}{8} + 1 = 2$$

Density of metal $\rho = \dfrac{2A3\sqrt{3}}{6 \times 10^{23}.8d^3}$ g/cm^3

$$d = \sqrt[3]{\frac{\sqrt{3}A \times 10^{-23}}{8\rho}} = 1{\cdot}294 \times 10^{-8} \sqrt[3]{\frac{A}{\rho}}$$

For sodium $A = 23$ and $\rho = 0{\cdot}97$ g/cm^3

$$\alpha = 1{\cdot}294 \times 10^{-8} \sqrt[3]{\frac{23}{0{\cdot}97}}$$

$$= 3{\cdot}72 \times 10^{-10} \text{ m.}$$

That is the average internuclear distance between sodium atoms is 0·372 nm (3·72 Å) which is the metallic diameter of sodium and all other metals have radii in the same order of magnitude. The percentage free space occupied by the atoms in a body-centred structure is calculated by an exactly equivalent way as above.

$$\text{percentage occupied volume} = \frac{\text{volume occupied by metal atoms} \times 100}{\text{volume per unit cell}}$$

$$= 200 \times \frac{4}{3} \pi \left(\frac{d}{2}\right)^3 \div \frac{8d^3}{3\sqrt{3}} = 68 \text{ per cent}$$

So that there is 32 per cent free space which is more than 8 per cent than that in the face centred structure, because the body centred structure is not a close packed one.

In hexagonal close packed structures the unit cell is a rhomboidal right prism as illustrated and in order to calculate the area of the cell we note on Fig. 6.4b that:

$$\text{Area of base} = \frac{\sqrt{3}}{2} d^2, \text{ therefore: height} = 2x.$$

To obtain x in terms of d we note that in $\triangle\text{DEF}$

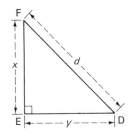

$$x^2 = d^2 - y^2$$

$$y = \frac{2}{3}\frac{\sqrt{3}}{2} d = \frac{d}{\sqrt{3}}$$

$$x^2 = d^2 - \left(\frac{d}{\sqrt{3}}\right)^2 = \frac{2d^2}{3}$$

$$x = \sqrt{\frac{2}{3}} \times d$$

But height $= 2x = 2d\sqrt{\dfrac{2}{3}}$

Volume of the cell $=$ Area of base $\times$ height

$$= \sqrt{3}\frac{d^2}{2} \cdot 2d\frac{\sqrt{2}}{3} = \sqrt{2}d^3$$

There are 6 atoms at corners shared by 6 unit cells, therefore average per unit cell $= 1$

There are 3 atoms in middle layer shared by 3 unit cells, therefore average per unit cell $= 1$.

Total number of atoms per unit cell $= 2$.

$$\text{Density of metal, } \rho = \frac{2A}{6 \times 10^{23}\sqrt{2}d^3} \quad \text{g/cm}^3 \quad A = \text{atomic weight}$$

From which

$$d = 3\sqrt{\frac{(A \times 10^{-23})}{(3\sqrt{2}\rho)}} \quad \text{cm}$$

$$= 1 \cdot 335 \times 10^{-8}\, 3\sqrt{\frac{A}{\rho}} \quad \text{cm}$$

144

Zinc   Ideal

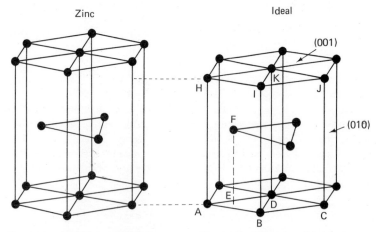

Fig. 6.4(a). The close packed hexa-   Fig. 6.4(b). The unit cell in the hexagonal structure of zinc is slightly more   gonal close packed structure
open than the normal close packed
lattice (the ratio of the axes is 1·856:
1·633)

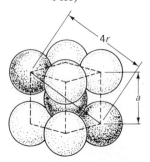

Fig. 6.4(c) The atoms in a body-   Fig. 6.4(d). The unit cells with atoms of
centred cubic structure   different sizes

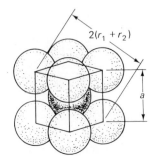

For magnesium which has a hexagonal structure

$$A = 24\cdot3 \quad \rho = 1\cdot74 \text{ g/cm}^3$$

$$d = 3\sqrt{\frac{(24\cdot3 \times 10^{-23})}{(3\sqrt{2} \times 1\cdot74)}}$$

$$= 0\cdot321 \text{ nm}$$

Therefore the average internuclear distance between magnesium atoms is
0·321 nm.

$$\text{Percentage occupied space} = 2 \times \frac{4}{3}\pi\left(\frac{d}{2}\right)^3 \times 100 \div \sqrt{2}d^3$$

$$= 74 \text{ per cent}$$

In the hexagonal close packed structure there is 24 per cent free space—the same as the cubic close packed.

In the Periodic Table the first two groups, that is the alkali and the alkaline earth metals, are regarded as true metals. In addition, transition metals such as nickel, palladium and platinum are also regarded as true metals.

Plate 6.1. Beryllium

Metals in higher group numbers however have reducing amounts of metallic character. Even within the first two groups the metallic character is considered to increase with atomic number. Thus beryllium has not an ideal close packed structure (Plate 6.1). The structure is distorted and some of the internuclear distances are longer than required by the ideal hexagonal close packed structure. Zinc and cadmium have distorted close packed hexagonal structures in which the axial ratios are about 1·87 rather than 1·63. Mercury is more 'normal' in that it is trigonal, each atom having six close neighbours. In Group 3, that is boron, aluminium, gallium, indium and thallium, boron is a 'metalloid'. Aluminium has the face-centred cubic structure, α thallium the close packed hexagonal, but indium has a slightly distorted face-centred cubic structure and gallium is unusual in that each atom has a coordination of 7. In Group 4, that is carbon, silicon, germanium, tin and lead, the metallic character of the elements increases with atomic number. Thus carbon and silicon are essentially non-metallic; germanium is a metalloid; tin and lead are truly metallic although their oxides, especially those of tin are amphoteric. As typical of the increasing covalent character the strength of the single interatomic covalent bond decreases from carbon to tin. Carbon, silicon, germanium and grey tin all have the diamond structure in which every atom is tetrahedrally bonded to four neighbours. The carbon C—C bond strength is 350 kJ mol^{-1} while the Sn—Sn bond strength is only 155 kJ mol^{-1}. The bond energy is defined as the energy required to separate the atoms in the required molecule or metal in its standard state to infinity. In typical covalent compounds, that is in non-metallic compounds, the electrons are essentially localised

146

between the atoms. In metals, however, the electrons are not localised between the atoms and grey tin and germanium are semi conductors since they do possess some electrical conductivity at elevated temperatures when the valency electrons are not completely localised. The electrical conductivity is much less in silicon than in true metals. Lead has the typical metallic face-centred cubic structure.

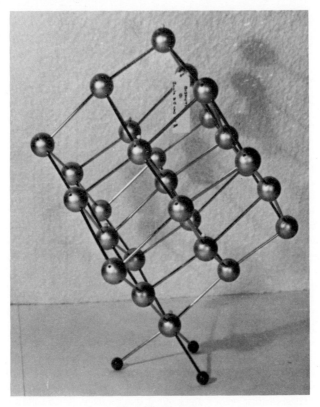

Plate 6.2. Bismuth

In Group 5, which contains nitrogen, phosphorus, arsenic, antimony and bismuth (Plate 6.2), nitrogen and phosphorus are strictly non-metals (see Chapter 5). The structures of arsenic, antimony and bismuth are built up from puckered sheets of atoms but the bonding between each atom is not strictly that of a typical metallic structure. In any layer of atoms each one has three covalently bonded neighbours, and each layer is effectively a giant molecule which is held to neighbouring sheets by a weak interlayer force. This structure is reminiscent of the structure of graphite and the layers are able to move freely over one another. The bond angle between the atoms within one sheet is always approximately 100°, which emphasises some directivity or covalent character of the bonds.

Alloys

There are two types of solid solutions, the substitutional and interstitial. Substitutional alloys are obtained by starting with one element and gradually substituting the atoms of that element with atoms of a different element. Gold and silver have the face-centred cubic structure, both gold and silver are univalent elements, and their atoms have very similar atomic radii of about 0·14 nm. Gold and silver form a series of continuous substitutional solid solutions which each have face-centred cubic structures, the different atoms being distributed at random among the different lattice sites. Such continuous solid solutions are obtained when the atomic radii of the elements do not vary by more than 14 per cent and provided the valencies of the elements are similar.

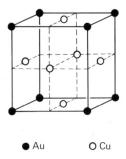

● Au   O Cu

Fig. 6.5. The gold-copper alloy (AuCu₃)

This alloy contains 25 per cent of the total number of moles of gold and has a cubic face-centred structure in which the gold atoms form the corners of the squares and the copper atoms are at the centres of the faces

The gold-copper system is a borderline case, the gold atom is 14 per cent larger than the copper atom and consequently at very high temperatures, which favour a random distribution of the elements, there is a solid solution. If this solid solution is suddenly cooled or quenched then the cooled material maintains the structure of the solid solution. If, however, the hot solid solution is slowly cooled or annealed, the gold and copper atoms separate out, with the result of alternate layers of gold and copper atoms when the composition of the alloy is AuCu, but the structure has no longer a cubic lattice. When the alloy AuCu is quenched the crystal is cubic and a is 0·386 nm, but when the alloy is annealed there are two parameters where a is 0·395 nm and c is 0·37 nm. When the alloy of composition AuCu₃ is annealed the gold atoms form the corners of the cubes, and the copper is at the face-centres of the cube (Fig. 6.5).

In a mixed disordered crystal of any kind (not only of metals, it could be of say anthroquinone and anthrone, discussed at the end of this book) which is based on a lattice having primitive translations a, b, c, and some process such as ageing or annealing transforms it into an ordered structure which is based on a lattice which has primitive translations $2a$, b, $4c$, the latter is called a superlattice structure of the former. The primitive translations of the superlattice are integral values of the smaller lattice. The word superlattice is a bad name but it is in common usage. Mathematically the lattice with the smaller translations is called a sublattice of the former or parent lattice as all the points are equivalent in the parent.

Ordered alloys then are obtained by annealing and disordered alloys are obtained by quenching. The properties of the ordered and disordered alloys often differ considerably even though the average composition of the alloys are the same. In the disordered alloy $AuCu_3$ the planes are unable to move over one another, as if the planes are joined together. The ordered alloy has a structure closer to that of the pure metal and is more ductile, more malleable and has a higher electrical and thermal conductivity. In some cases there is a definite temperature at which there is equilibrium between the ordered and disordered structures. The temperature is referred to as the critical order–disorder transition temperature and is often characterised by the fact that at this temperature the metal has a high specific heat, for it is at this temperature that atoms are being disordered, or in thermodynamic terms the entropy is being increased, a process which requires energy. In some ways this change may be regarded as melting, because at the melting point of a substance there is a typical change over, often from virtually complete order to a much less ordered system. In cobalt, which normally has the hexagonal close packed structure, there is an abrupt change at the critical *allotropic* transition temperature from the regular hexagonal structure to one in which eleven layers have the hexagonal structure and the next eleven have the close packed cubic structure.

The alloy AuCd is isostructural with NiTi (nitinol) which has many unusual properties for it can exist in four different forms. The first has a simple body-centred cubic structure being stable above 948 K (675 C) and the second has a caesium chloride structure being stable between 438 K (165 C) and 948 K (675 C). The third is a mixed structure resulting from a transformation below 438 K and the fourth which has a complex structure is stable below 373 K. The reason why this alloy is unusual is that a wire of the material which is coiled at room temperature will uncoil suddenly at 438 K when the structure changes. On cooling the wire recoils as the disordered structure is reformed. If you feel like doing this experiment a flat disc will do the trick but notice how quickly the change occurs as the transition temperature is characteristic of this particular alloy. What happens if the alloy is coiled above 438 K and straightened the first time it is cooled?

Disorder, which is quite distinct from a dislocation or deformation or from impurities, is usually characterised by giving a diffuse X-ray scattering. Calcium oxalate $(CaC_2O_4.2H_2O)$ may lose a molecule of water when the substance is ground and one or more of the X-ray diffraction rings becomes diffuse.

When a true metal is alloyed with an element of Group 5 or 6 the range of solid solution is much more restricted than if a metal of Group 1 or 2 were used. The resultant may be metallic in appearance and may conduct electricity but often the electrical conductivity is minimal at some simple ratio of the atoms. It is therefore likely that this type of alloy constitutes the formation of a definite chemical compound. The usual structure of this type of alloy is that of nickel arsenide. The sulphides, selenides, tellurides, arsenides, antimonides and bismuthides of many transition metals have this structure. The structures are examples of defect structures and in a typical example, CrS, the holes left by the absent chromium ions are often ordered and in effect form a superlattice. The incomplete defect structure is defined as that in which equivalent sites in a structure are only partially occupied, leaving vacancies within the structure. Mixed crystals are sometimes examples of defect structures with complete structures. In a mixed crystal of potassium chloride and potassium bromide the bromide and chloride ions are distributed among equivalent sites within the sodium chloride type structure. The iron sulphide pyrrhotite FeS is always deficient in iron and has an incomplete defect structure. Silver iodide has three structures but the alpha form, which is stable above 423 K has a body-centred cubic arrangement of the iodide in which the silver ions are able to move among the fixed iodide ions. The silver ions are therefore distributed at random amongst the fixed iodide structure, which finally collapses above 823 K.

The compound Ag_2HgI_4 is an example of a substance with a combination of complete and incomplete substructure. There are two structures, the alpha which is stable above 323 K, and the beta form which is stable below 323 K. The iodide ions form face-centred cubic lattices in both structures, but in the beta form the silver and the mercury cations occupy definite sites in the crystal lattice. Both the silver and the mercury ions are in exactly equivalent sites, being surrounded tetrahedrally by four iodide ions. In the alpha form the cations are evenly distributed over four different types of sites. It is possible to regard this structure as that of zinc blende in which the metal ions sites are accommodating atoms of two types and in addition some metal sites are unoccupied.

Interstitial compounds are formed when small atoms are able to pass into the holes or interstices within the metallic structure. Earlier in the chapter we saw that in the close packed structures there was about 25 per cent free space. Atoms such as hydrogen, boron, carbon and nitrogen are able to fill these holes and in some cases drastically alter the properties of the metal. Steels are essentially iron into which carbon has been added and are always much harder than the parent ion. Molybdenum nitride (MoN) and tungsten carbide (WC) are as hard as diamonds. In practice an interstitial alloy is formed provided the radius ratio of the small element to the metal is less than 0·59. When the radius ratio is above this value the structure of the metal becomes distorted. In close-packed arrangements there are two types of holes, the larger ones are formed by six metal atoms and are octahedral and on average there is one octahedral hole per metal atom. The

150

smaller holes are tetrahedral and there are twice as many tetrahedral holes as octahedral holes. Austenite is one of the many phases of the iron-carbon system, and has a face-centred cubic (close packed) atom iron arrangement and the carbon atoms occupy some of the octahedral holes. The unusual fact is that in all the phases of either manganese or iron or cobalt or nickel with nitrogen the nitrogen is replaceable atom by atom by carbon. This leaves us to wonder what the valency forces in interstitial alloys are like, but as yet there are no satisfactory explanations.

Dislocations

Dislocations are the boundaries between the parts of a plane which has moved and the part of a plane which has not moved. Metals and most of the other crystalline solids may deform by planes gliding over one another. Calculations show, however, that the stress needed to cause one layer of atoms to move over another layer is about one thousand times less than the observed values. This discrepancy between the calculated and the observed values is explained by the fact that planes do not move over one another completely but move over one another little by little as the movement of the

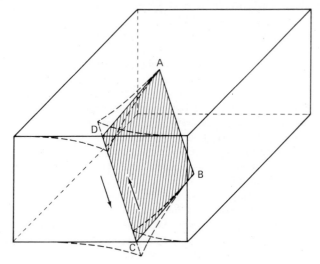

Fig. 6.6. Dislocation

above dislocations (Fig. 6.6). The reader may make a model of a dislocation for himself by taking a piece of rubber and cutting a plane within the rubber and if the two surfaces of the slit are moved in relation to one another, the region of strain along AB is then similar to the dislocation. Near dislocations the structure of the crystal is distorted (the regular periodic arrangement of the atoms in a perfect crystal is however, seldom achieved). Real crystals possess various kinds of imperfections (Fig. 6.7), dislocations are just one example of an imperfection in a crystal. Grain boundaries are formed when two different regions of crystal orientation have grown independent of one another and then meet. In Chapter 2 the grain boundaries on the etched surface of aluminium (Plate 2.4) were clearly seen. In Plate 6.3 a model of a dislocation is illustrated and plate 6.4 shows how a dislocation moves through a metal when a stress is applied.

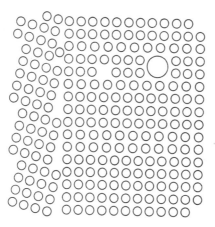

Fig. 6.7. Disorder in a monatic cubic crystal—diagrammatic (showing a grain boundary, dislocation, impurity atom, and a vacant site)

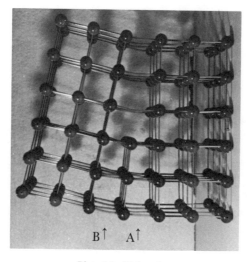

Plate 6.3. Dislocation

Plate 6.4. Movement of dislocation

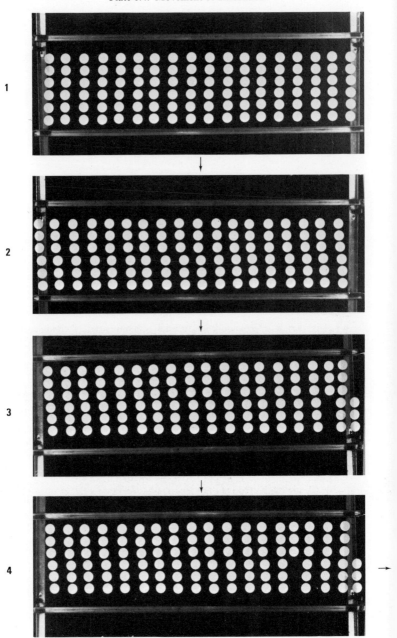

153

Plate 6.4 (*cont:*)

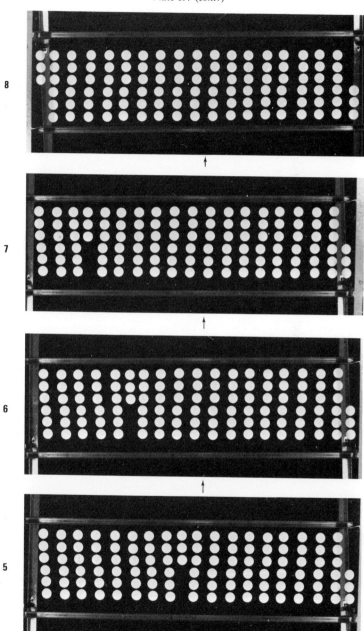

Crystals may be built up from a screw dislocation (Fig. 6.8), as one layer of a crystal grows, irrespective of whether it is an alloy or one of the crystals we met in other chapters, an imperfection may be formed within that layer. The remaining layers fill up to minimise the number of imperfections within the earlier layers and the layer may be slightly displaced at a small angle to the lower layers. As the layers build up they do so in the formation of a spiral.

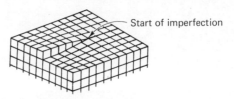

1. The screw dislocation

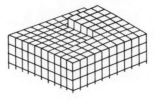

2. The layers of ions or molecules build up

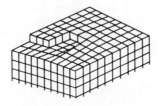

3. The first stage of the spiral

Fig. 6.8. The formation of a spiral from a screw dislocation

Electrical conductivity and semi-conductors

So far we have not explained how it is that metals and alloys conduct electricity. Within each metal it is best to regard the structure as being composed of ions between which electrons freely move in a random manner. When a potential is applied to the metal the electrons move from a region of high potential energy to a lower energy. Semi-conducting materials are those which are insulators at low temperatures but conduct electricity at high temperatures and are used in such things as transistors. The usual interpretation for conductors and semi-conducting materials is given by the Band Theory. We have seen how in covalent compounds the electrons which are waves spend more time between the nuclei and tend to be localised between atoms but in metals the electrons flow between the ions. In isolated atoms the electrons occupy definite energy levels but when the atoms are brought together a metal structure is formed, the energy levels separate out into bands (Fig. 6.9). In conducting materials the bands overlap as in

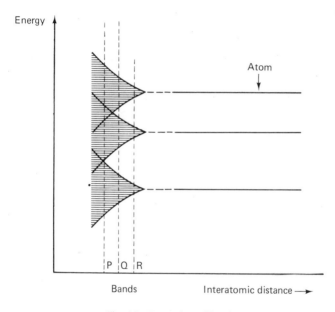

Fig. 6.9. Formation of bands

position P. In semi-conducting (Q) and non-conducting materials (R) the bands do not overlap. In conducting materials (Fig. 6.10) the uppermost electrons may flow from a high energy level to a low energy level. In the lower energy level however the level is completely filled and electrons are unable to flow, with the resulting lack of flow of electricity. A usable analogy to this filled band is the fact that the main bulk of water in a completely filled tank is unable to move (the molecules are, however, in motion). By analogy to the partially filled band water in a partially filled tank is able to move when the tank is tipped. In insulating materials the energy difference between the completely filled band, or valence band, and the conduction band is wide and electrons are unable to pass from the insulating band to the conduction band. In semi-conducting materials the electrons are able to move from the valence band to the conduction band when they have suffi-cient energy to jump the gap. At low temperatures the thermal energy of the electrons is too small to allow this jump, but at high temperatures the thermal energy is sufficient to allow the electrons to jump. There are two types of semi-conductors, namely intrinsic and extrinsic. The extrinsic type is the one which really concerns us mostly as the extrinsic semi-conductors arise

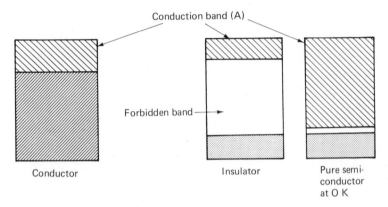

Fig. 6.10. Bands in a conductor, insulator and semi-conductor

from defects in the semi-conductor crystal structure. These structure defects are generally structure vacancies, that is either the absence of a cation or anion, or impurity ions may be added into the host structure. When there is an excess of negative charges the semi-conductor is an N-type, while if there is an excess of positive charges the semi-conductor is called a P-type (positive charge carriers). The electric current may be carried by the negative or positive charges, and in general the negative electrons travel faster than the positive charges.

The defects within the crystal structure often give rise to energy levels between the main insulator and the main conduction band. Levels which are introduced near the conduction band are described as 'donor levels' from which an electron may easily be excited into the conduction band. Levels which occur near the valence or insulator band are described as 'acceptor levels', that is they accept an electron from the valence band, thereby producing a hole in the valence band. The structure of the non-stoichiometric compound titanium oxide is a typical example of a defect structure (Fig. 6.11) and the defects may be caused by an absence of titanium ions. The result is that there is a positive hole which is trapped within the main structure, this would give a P type of semi-conductor. It is quite possible that there would be an absence of an oxide anion resulting in the fact that electrons are trapped within the main lattice and if this happened a great deal the result would be an N-type of semi-conductor. One question we might ask is 'Is it then possible to make N-type and P-type semi-conductors to order'? The brief answer is that the main requirement for N-type of semi-conductor would be to introduce an impurity, e.g. arsenic, into a main host lattice, e.g. germanium. On average arsenic atoms have one more electron

Ti^{2+}	O^{2-}	Ti^{2+}	O^{2-}	Ti^{2+}	O^{2-}
O^{2-}	Ti^{2+}	O^{2-}	Ti^{2+}	O^{2-}	Ti^{2+}
Ti^{2+}	O^{2-}	(2p)	O^{2-}	Ti^{2+}	O^{2-}
O^{2-}	Ti^{2+}	O^{2-}	Ti^{2+}	(2e)	Ti^{2+}
Ti^{2+}	O^{2-}	Ti^{2+}	O^{2-}	Ti^{2+}	O^{2-}

Electrons (e) are trapped in anion (O^{2-}) vacancies.

Positive holes (p) are trapped in cation (Ti^{2+}) vacancies

Fig. 6.11. The structure of non-stoichiometric TiO (TiO$_{1.35}$–TiO$_{0.6}$)

than the germanium and the result is that there are negative electrons (holes) trapped within the structure. The underlying theory for making these semi-conductors is quite easy because arsenic is in group 5b and has an atomic number of one more than germanium and consequently neutral arsenic has one more electron than neutral germanium. Gallium has an atomic number of one less than germanium and when gallium is introduced into a germanium lattice as an impurity there is an excess of positive holes, thus making a P-type of semi-conductor. A P–N junction may act as a rectifier in an a.c. circuit. A rectifier acts so that current is able to move in one direction but not in another. In Fig. 6.12 a water bottle analogy is made for

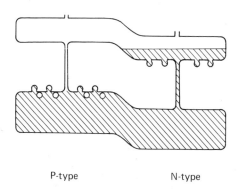

P-type   N-type

Fig. 6.12(*a*). Water-bottle analogy for a P–N junction

a P–N junction. The junction occurs at the intersection between the P- and N-type regions, the main insulator band being at the base and the conduction band at the top. Note that in the N-type the conduction band is partially filled but normally in the P-type of semi-conductor it is vacant. In the N-type the valency band is completely filled but in the P-type small holes or bubbles represent the positive holes within the lattice. The small recesses just above the valency band in the P-type represent the negative charges which balance the positive holes or bubbles. When the whole apparatus is tipped up such that the N-type is above the P-type, water may flow from the

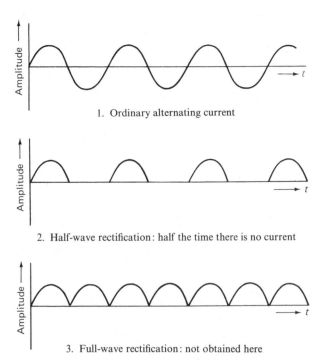

1. Ordinary alternating current

2. Half-wave rectification: half the time there is no current

3. Full-wave rectification: not obtained here

Fig. 6.12(b). Half-wave rectification obtained by a single transistor or diode

N-type to the P-type, and the bubbles move from P-type to the N-type. That is, there is an overall transfer of electric current by movement N-type. That is, there is an overall transfer of electric current by movement of water and bubbles, that is by movement of negative and positive charge. If the apparatus is now tipped so that the P-type is above the N-type then the bubbles are unable to move to the N-type and the water is unable to move to the P-type, and there is no resultant movement of electric current. Therefore alternate applications of potential in an alternating electric current on passing through an N–P-type junction only does so for one particular bias of the cycle. Hence the alternating current is changed to a direct current, giving a half-wave rectification (see Fig. 6.12b).

The piezo-electric effect

While we are discussing electric effects in crystals it is interesting to ask the question 'Why is it that crystals can be used in gramophone pick-up arms?' The reason is that the distribution of charge in these crystals is asymmetrical. The crystals do not have centres of symmetry and when the crystal is physically distorted by moving over the groove in the gramophone record, then the ions within the crystal are slightly moved with respect to one another. The result is that the positive and negative ions are slightly separated and since electricity means a flow or separation of electrons from positive charges there is a resultant flow of electricity. Lithium sulphate monohydrate is a typical example of a piezo-electric crystal (Plate 6.5). Within the

Plate 6.5. Piezo-electric crystal lithium sulphate ($LiSO_4 \cdot H_2O$)

crystal there are layers of sulphate groups which are widely separated from the next one. When the crystal is compressed the average charge distribution is altered and there is a similar effect when the crystal reforms its original shape. The piezo-electric effect therefore arises from the physical movement of ions which results in a flow of electricity.

The phenomenon of piezo-electricity is defined as 'the production of electricity by mechanical strains in crystals belonging to certain classes'. The polarisation is proportional to the strain applied and changes sign with the strain. If strain is applied to crystals of zinc blende, sodium chlorate, tourmaline, quartz, topaz, calamine, tartaric acid, cane sugar and rochelle salt, small deflections are seen on an electrometer which is connected to tinfoil electrodes placed correctly along certain orientations. Piezo-electric deformations of the crystals are directly proportional to the electric field and the deformations are reversed when the sign of the electric field is reversed. The mechanics of the piezo-electric effect is not completely clear but it has been found to be closely related to the thirty-two crystal classes. Piezo-electric crystals were used during the war in 'Echo method detectors'. Langevin used piezo-electric crystal plates of quartz as a source of very high frequency and started the new field of Ultrasonics. Piezo-electric crystals are used in a very accurate quartz clock in which a vibrating quartz crystal replaces the more traditional swinging pendulum, and the resulting clock is more accurate than many earlier astronomical clocks. Indeed piezo-electric crystals have certain peculiarities, when subjected to a frequency by mechanical means, which are used in stabilising and filtering electric current. Rochelle salt has a very high piezo-electric effect and this effect is used in microphones, telephone receivers and record cutters.

The field-ion microscope

So far we have only mentioned in rather vague terms about atoms being formed in layers and the main evidence we have talked about has been the electron microscope. The field-ion microscope provides one of the first really concrete answers to the questions 'Have we ever seen atoms?' The answer is now that we are able to see in detail the arrangements of atoms on the surface of metals such as tungsten. The main principle of the method is that there is a maximum of electrostatic charge where the curved surface has its smallest radius. Also it is possible to have a very high charge on the surface of a finely pointed metal tip. The specimen is a sharply pointed end of wire which faces a fluorescent screen in a vacuum chamber. The specimen is maintained at a high positive potential with respect to the fluorescent screen with the result that there is an intense electric field at the sharply curved surface. A gas, which is called the image gas, is admitted under low pressure to the vacuum chamber and the gaseous atoms are ionised by the loss of an electron to the positive charge on the metal tip. The resultant positive ions are repelled very strongly from the positive tip in directions which are normal to tangents to the surface at the metal tip. The ions form an image on the fluorescent screen and a very highly magnified image of the tip is seen. The magnification is the distance of the image to the specimen divided by the radius of the specimen point. There are therefore two advantages obtained by making the tip with a very fine radius, the first is to obtain a high charge and the second is to increase the magnification. Magnifications of more than a million are quite normal with applied voltages of 25 kilovolts (kV) and the field strengths are up to 5×10^{10} volts per mm. The results of the field-ion microscope are seen in Plate 6.6. The dots which are clearly seen represent ions which have come from individual atoms. Close examinations of certain regions of the micrograph show that, as we have said before, atoms in a given layer are surrounded by six other atoms. The rings represent sections across different crystal faces. Field-ion microscopy also provides definite evidence that hydrogen is absorbed onto

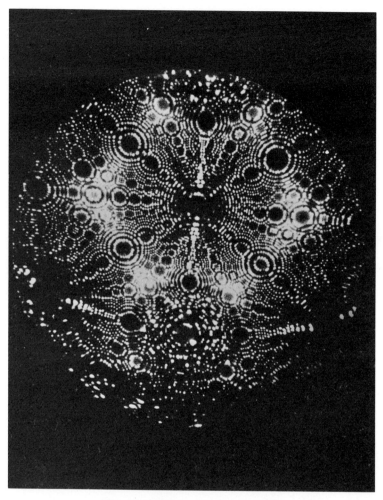

Plate 6.6. Field-ion emission micrograph

the surface of the metals when they act as catalysts. Tungsten acts as a catalyst in a variety of hydrogenation reactions and if the field emission micrograph of tungsten with helium as the image gas is compared with that with hydrogen as a carrier gas the emission micrographs are quite different. The reason is that the hydrogen passes into the interstices between the tungsten atoms probably to form the interstitial hydrides, which were discussed earlier in this chapter. The field-ion microscope is still being studied and developed as a new technique. It is hoped that in the future it will provide evidence not only of the structures of metals and alloys themselves but more definite evidence for the reactions of metals which take place on the surface of the metal. Metals act as catalysts because they are able either to absorb physically or form a chemical bond with the reactants in a given reaction. For example in the hydrogenation of ethylene

$$C_2H_4 + H_2 = C_2H_6$$

is catalysed by metals such as nickel, palladium and platinum. Evidence so far shows that when ethylene is introduced into the field-ion microscope the surface structure of the tungsten is drastically altered, but care must be taken in the interpretation of the micrographs because the absorption of molecules onto the surface will affect the way in which positive ions of the carrier gas are formed. It is possible that biological molecules may be studied with the field-ion microscope, and the aim is to deposit a single biological molecule under definite controlled conditions onto the rod in the field-ion microscope, so as to obtain an image which will reveal the atomic and molecular arrangements of the biological molecule. The main difficulties regarding the structure of biological molecules are discussed in Chapter 7.

Masers

Physicists and electrical engineers were frequently beset with problems of how to deal with electric currents with very small amplitudes. The problem was solved by the introduction of masers, which is an abbreviation for Micro-wave Amplification by Stimulated Emission of Radiation. The maser usually consists of a ruby, neodymium or calcium fluoride crystal, doped with atoms or ions, which are themselves able to give a maser action and energy changes occur in the impurity atoms such as chromium ions. The method of working is designed such that energy is absorbed so that electrons are excited from a lower energy level to a higher energy level with absorption of energy $\triangle E$ (see Chapter 3) which corresponds to a frequency v. When electrons transfer from a higher energy level to a lower energy level there is an emission of energy of frequency v. The ruby crystal is cooled with liquid helium so that all of the electrons are in the lower energy state. A radio frequency enters the ruby crystal and sets up a strong microwave field so that electrons are excited to the higher energy level in which the electrons are maintained until they are triggered to go to the lower energy level by a weak input signal such as that from a satellite. The output is of the same frequency as the weak input signal but is much amplified. The main purpose of masers is, therefore, as low noise amplifiers capable of detecting signals from satellites.

Lasers

The word laser is an abbreviation of Light Amplification by Stimulated Emission of Radiation. Crystal and gas lasers are used to convert a very high energy flash of light into a coherent radio signal. Some ruby laser crystals are shown to the left-hand side of Plate 6.7 and some neodymium

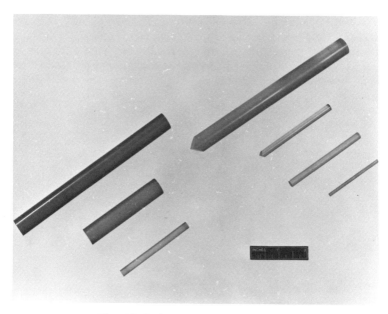

Plate 6.7. Ruby and neodymium laser crystals

Plate 6.8. Concentrated laser beam vaporising a small carbon plate

164

ones to the right hand side. The main feature of lasers is that light of energy
B (Fig. 6.13) is spread over a time t_1 to t_2 which is longer than the time that
the stimulated frequency is emitted. The total energy input is the same as
output but the emitted beam is very intense. Light from a cylindrical flash
lamp whose centre is concentric with the cylindrical ruby crystal passes

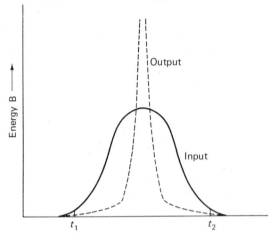

Fig. 6.13. Lasers
Total input energy ⩾ total output energy

into the crystal and excites electrons to higher energy levels. The light is then
reflected from end to end of the crystal which is 100 per cent silvered on one
end and partially silvered at the other from which the beam eventually
emerges and the energy emerges as a single intense beam of radio-frequency
waves. The main advantages are that the output is strong, occurring over a
very narrow range of frequencies, and may be focused on a very small spot.
The main use of lasers is in space research as the signal may be sent over
large distances but they are becoming increasingly important in atomic
physics and localised surgery. Weapons research departments are developing crystal lasers to be used as direction finders but the use of lasers as
ray guns is not feasible at present as the beam is not sufficiently intense.
The emergent ray can have sufficiently high energy (Plate 6.8) to cause a
small plate of carbon to vaporise.

Electro-optical modulators

Modulation means altering or affecting a particular property, thus a frequency modulated source is one in which the frequency is altered. Crystal
modulators use crystals of ammonium dihydrogen phosphate (ADP),
potassium dihydrogen phosphate (KDP), potassium deuterium phosphate
(KD*P) whose formation was discussed in chapter 1. The crystals which

1a	2a	3a	4a	5a	6a	7a	8	8	8	1b	2b	3b	4	5b	6b	7b	0
H mol																	He [cubic]
Li body hex	Be hex											B* complex 3 dim	C diamond (graphite)	N mol	O mol dioxygen	F mol	Ne [cubic]
Na body	Mg hex											Al* distorted c.p.	Si diamond	P w-P_4 b-layer	S rh. mono plastic S_6	Cl γ mol	A cubic
K body	Ca cubic; body; hex	Sc cubic; hex	Ti body; hex	V body	Cr body	Mn* complex	Fe γ-cubic α-body hex	Co cubic hex	Ni cubic hex	Cu cubic ε-hex	Zn* distorted c.p.	Ga* distorted c.p.	Ge diamond	As y-As_4 g-layer	Se helical chains	Br mol	Kr [cubic]
Rb body	Sr cubic; body; hex	Y hex	Zr body α-hex	Nb body	Mo body hex	Tc body	Ru hex	Rh cubic	Pd cubic	Ag cubic	Cd* distorted c.p.	In* distorted c.p.	Sn* g-diamond w-complex	Sb layer	Te helical chains	I mol	Xe [cubic]
Cs body	Ba body	La† cubic	Hf body; hex	Ta body hex	W body	Re body	Os hex	Ir cubic	Pt cubic	Au cubic	Hg* complex	Tl* distorted c.p.	Pb cubic	Bi layer	Po simple cubic	At [mol]	Rn [cubic]
Fr [body]	Ra‡ [body]																

(Transition series spans groups 3a–8 in periods 4–6.)

f-block series:

		3a	4a	5a	6a	7a	8	8	8	1b	2b	3b	4	5b	6b
		†Ce cubic hex	Pr	Nd	Pm	Sm	Eu body	Gd hex	Tb hex	Dy hex	Ho hex	Er hex	Tm hex	Yb cub	Lu hex
		‡Ac cubic	Th cubic	Pa* complex	U* body	Np* body	Pu* body cubic	Am* complex	Cm	Bk	Cp				

(Pr → Sm: ←—— complex structures ——→)

Fig. 6.14. Summary of main structures of some elements in the solid state

Key: mol = molecular; body = body centred cubic; hex = hexagonal close packed; cubic = face centred close packed; 3 dim = 3 dimensional; [cubic] = possibly cubic; distorted c.p. = distorted close packed; w = white; b = black; y = yellow; g = grey; tr = trigonal (orthorhombic); mono = monoclinic; * = complex.

are in the form of flat plates are placed between two transparent electrodes which may be made from vacuum deposited gold on glass which allows a maximum of light to pass. A minimum amount of light is allowed to pass through the modulator when no electric field is applied. When an electric potential is applied the crystal becomes biaxial (whereas normally it is uniaxial) and the incident plane polarised beam is split into its two mutually perpendicular components and the result is an elliptically polarised light beam. The eccentricity of the elliptically polarised light varies with the applied potential and generally the higher the voltage the higher the intensity of light which is therefore modulated by the applied voltage.

Questions

1 Write down and draw the structures of sodium, one form of iron, aluminium and two forms of tin.

2 What is meant by the term allotrope? Discuss the allotropes of two metals.

3 What are the characteristic features of metals? How are these features explained by the structure of metals?

4 Distinguish clearly between (quoting examples in each case): (a) dislocation; (b) deformation; (c) defect lattice.

5 What is an alloy? Are all alloys solid solutions? How do substitutional and interstitial alloys differ?

6 What is meant by the term critical order–disorder temperature? Give clear examples to illustrate your answer.

7 How do metals act as catalysts? Why are some metals better catalysts than others? Why are hydrogenation reactions particularly well catalysed?

8 Discuss the statement 'Metals have ordered structures whereas non-metals have disordered structures'.

9 Discuss the important features of the field-ion microscope. How does this compare with the electron microscope? Give reasons for which of the two methods you would use, if any, to study the following:
(a) flat crystal of tungsten; (b) needle of sulphur; (c) needle of iron; (d) biological molecules; (e) the reaction between ethylene and hydrogen catalysed by tungsten; (f) steels of varying composition.

10 What are the fundamental structures of metals? How do these structures explain the properties of metals? Why does pure copper give a diffuse X-ray diffraction pattern when it is subjected to a stress?

11 Estimate the Avogadro number given that for sodium $d = 3.72 \times 10^{-10}$ m and $\rho = 0.97$ Mg/m^3 (g/cm^3) and that sodium has a body-centred structure.

12 How are metallic radii measured? Are they constant for a given element?

13 The following statements are meant to provoke discussion:
  (a) The compound AuCd is an interstitial compound of cadmium.
  (b) Because potassium and argon are both metals they both have the cubic close packed structure.
  (c) The simple cubic structure is close packed.
  (d) Like sulphur, crystals of oxygen contain chains of oxygen atoms.

14 Discuss the following statements:
  (a) Structures of elements in a given group become more metallic as the atomic number of the element increases.
  (b) Grey tin has a diamond-like structure but white tin has a complex structure.
  (c) Yellow arsenic consists of As_4 units but grey arsenic has a layer structure.
  (d) Oxygen should be called dioxygen.
  (e) Orthorhombic and monoclinic sulphur contain S_8 rings but plastic sulphur contain S_6 units.
  (f) Iodine has a molecular structure but argon has a cubic close-packed structure.
  (g) Boron has a three-dimensional complex structure.
  (h) Iron can have several structures.
  (i) The crystal structure of solid hydrogen resembles that of chlorine rather than lithium.
  (j) Alkali metals all have the body-centred cubic structure.
  (k) Group 6a all have the body-centred cubic structure but in group 6b there is a change from molecular to the simple cubic structure.
  (l) Alloys are solid solutions.
  (m) The gold atom is 14 per cent larger than the copper atom.
  (n) The compound CrS has a defect superlattice.
  (o) Interstitial compounds are only formed by metals with the smallest elements and can sometimes retain the same properties as the parent metal.

References

AGAR, A. W., 'Developments in electron microscopy', *A.E.I. Research and Development Journal*, **5**, Sept. 1963.

AGAR, A. W., *British Journal of Applied Physics*, **11**, 1960, 185.

BENFEY, T., 'Geometry and chemical bonding', *Chemistry*, May 1967.

BERNAL, J. D., *The Problem of the Metallic State*, a general discussion held by the Faraday Society, March 1929, Butterworth Scientific Press.

DICKENS, P. G. and WHITTINGHAM, M. S., 'The Tungsten bronzes and related compounds, *Quarterly Review of the Chemical Society*, **22**, 1968, 30.

GEACH, G. A. and JONES, F. O., 'Transmission electron microscopy of dislocations in metals', *A.E.I. Engineering*, 1961.

168

LIVINSTON, R. L., 'The teaching of crystal geometry in the introductory course', *Journal of Chemical Education*, **44**, 1967, 376.

LONSDALE, K., 'Crystallography as a research tool in chemistry', *Chemistry and Industry*, 1966, 1154.

LONSDALE, K., 'Disorder in solids', *Chemistry*, **38**, 1965, 14.

WALLACE, C. A., 'Imperfections in single crystals direct observation by X-rays', *G.E.C. Journal of Science and Technology*, **32**, 1965, 63.

WELLER, P. F., 'An analogy for elementary band theory in solids', *Journal of Chemical Education*, **44**, 1967, 391.

7 Some organic and biological molecules

Introduction

There are many atoms in biological molecules but proteins consist of many simpler molecules which are joined together in an order which sometimes aids the three dimensional analysis. Proteins consist of amino-acids which are joined to give the peptide chain. There is one peptide chain per molecule of myoglobin and lysozyme, and four in one molecule of haemoglobin. These peptide chains, as we shall see, are asymmetrically distributed (apparently randomly) throughout the molecular structure, but a study of such molecules was only possible after early work in the study of organic molecules which was usually used to confirm the shapes predicted by chemists and to show how such molecules could be fitted into the measured unit cell. In 1921 Sir William Bragg studied the structures of anthracene and naphthalene but did not confirm the structures which are still written in text books of organic chemistry. Unfortunately W. H. Bragg thought that the carbon atoms in naphthalene and anthracene were in puckered hexagons. An Indian scientist, Banerjee, corrected the orientation of the X-ray beam and eventually J. M. Robertson showed that the correct molecular structures contained planes of carbon atoms. In 1929 Lonsdale determined the structure of hexamethylbenzene and proved that the benzene molecule was a planar hexagon. These two sets of studies had two important features. The first is that they were the first major studies of organic molecules and the second that new techniques were developed to determine the absolute intensities of the diffraction spots. Briefly, theory relates the integrated intensity when a crystal is steadily rotated through the reflection position to the intensity of the X-ray beam falling per second on the crystal. The integrated intensity for sodium chloride (400) is 1.09×10^{-4} (using molybdenum Kalpha radiation) which served as a basis for the determination of the absolute intensity for other crystals. In 1925 the Fourier analytical methods were developed as mathematical means of calculating the coordinates of the atoms in the unit cell from the position and intensity of the diffraction spots. The mathematics involves a great number of calculations but the phase problem is greatly simplified when there is a centre of symmetry, a set of isomorphous compounds with a variable atom or a replaceable heavy metal atom and the main simplification comes when the replaceable heavy metal atom is at the centre of symmetry. This last simplification gives us a clue to the answer of the question 'How is it that the structure of the phthalocyanines (Fig. 7.1), which are complex organic molecules, was determined before the structure of much simpler molecules?' The answer is that the phthalocyanines form isomorphous nickel (and copper) salts in which the metal atoms were at the centres of symmetry. The study of proteins by X-rays is simplified by isomorphous substitution of a heavy metal atom into the unit cell. There is one important difference, however, between the substitution in phthalocyanines and in proteins because substitution into proteins occurs in random sites and not into fixed, predetermined positions.

Recently several well deserved Nobel prizes have been awarded for the

Fig. 7.1. Copper Phthalocyanine (Monastral fast blue)

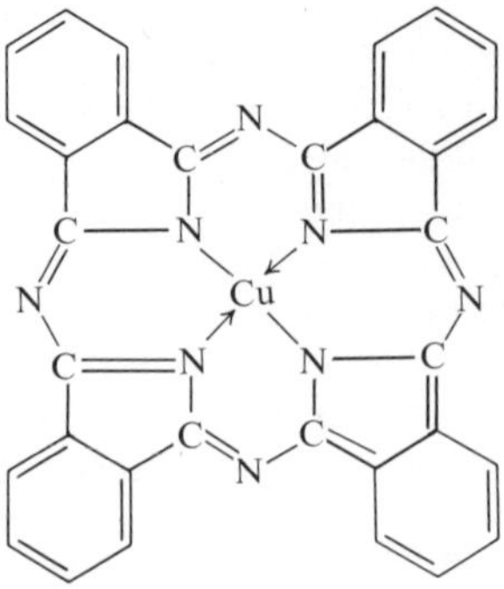

This molecule has a centre of symmetry and furthermore there is a heavy metal atom which may be changed for others at the centre of symmetry. Since the X-rays are scattered by electrons then as the atomic number, which is the number of electrons outside the nucleus, increases then the X-rays are scattered more. If the phase is such that the atom is making a positive contribution to the intensity then when the atomic number increases the intensity will increase. But if there is a negative contribution then the intensity will decrease. So that it is possible to find the sign of the phase.

---

study of some complex organic and biological molecules. The prizes included awards to Dr J. Kendrew for his work on the protein myoglobin, to Dr M. Perutz for his work on haemoglobin and Dr Dorothy Crowfoot-Hodgkin for her work on vitamin B_{12}. The work of these people together with the joint Nobel prize winners Crick, Watson and Wilkins for their work on DNA forms the initial basis for this chapter, in which viruses, penicillin and fibres are discussed and a word of caution is given.

Proteins

Proteins are found in all living things in an amazing variety of roles such as hair, muscle, cartilage, myoglobin and haemoglobin; in addition, chromosomes and viruses contain proteins and nucleic acids. The study of proteins is simplified because amino-acids are linked together in long chains called peptide chains. Despite their complex structure many proteins can be made to form crystals. Indeed objects as large as some viruses, with molecular weights of over 10 000 000, can be crystallised. The viruses or molecules in the crystals are arranged in regular ways just as the ions of sodium and chlorine are in salt crystals. In both myoglobin and haemoglobin the peptide chains are randomly distributed throughout the crystal and this randomness adds difficulties to an already complex problem. Moreover myoglobin and haemoglobin reversibly take up oxygen and it was important to know whether the two proteins took up oxygen in the same manner.

Amino-acids consist of a carboxyl group (COOH), which is an organic acid group and an amino group (NH_2) which is an organic basic group. Amino-acids therefore have both acidic and basic properties, and all have the same dipolar group (Fig. 7.2) and they combine together with the elimination of water to form the —NH.CHR.CO— group (where R is often an alkyl group) which is the skeleton of the polypeptide chain. Although over eighty amino-acids are known usually only twenty of the αamino-acids occur in proteins whose properties are determined by the sequence. The amino-acids differ by the nature of the side chain R, and specificity is provided by twenty different side chains whose first carbon atom is given the symbol beta and the carbon atom in the main chain is labelled alpha. The alpha carbon is attached to four different groups and is a centre of optical activity, there being two possible enantiomorphs but usually the L enantiomorph is found in natural proteins, whilst the D form occurs in bacterial cell walls. The hydrogen is regarded as being at the apex of a tetrahedron which is viewed from the top and the R, the amino and the carboxyl groups, succeed each other in a clockwise manner.

Of the better known amino-acids (Fig. 7.2) the simplest is glycine in which the side chain R is a H atom. In the next simplest ones, that is in alanine,

The Basic Structure

$$R-\underset{\underset{H}{|\alpha}}{\overset{\overset{NH_2}{|*}}{C}}-C\overset{\diagup\diagup O}{\diagdown O-H}$$

The Dipolar Group

$$R-\underset{\underset{H}{|}}{\overset{\overset{NH_3^+}{|}}{C}}-COO^-$$

The Repeating Unit in the Polypeptide Chain

$$-NH-\underset{\underset{R}{|}}{\overset{\overset{\alpha}{}}{CH}}-CO-$$

The Optically Active Form

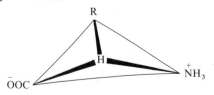

Fig. 7.2. Some commonly occurring amino-acids

Trivial name	Shortened name	Formula
Glycine	Gly	$$\underset{\text{H}-\underset{\displaystyle |}{\text{CH}}-\text{COOH}}{\overset{\displaystyle \text{NH}_2}{}}$$
Alanine	Ala	$$\underset{\text{CH}_3-\text{CH}-\text{COOH}}{\overset{\text{NH}_2}{}}$$
Valine	Val	$$\underset{\text{CH}_3\text{CH}-\text{CH}-\text{COOH}}{\overset{\text{CH}_3 \quad \text{NH}_2}{}}$$
Leucine	Leu	$$\underset{\text{CH}_3\text{CHCH}_2-\text{CH}-\text{COOH}}{\overset{\text{CH}_3 \qquad \text{NH}_2}{}}$$
Isoleucine	Ileu	$$\underset{\text{CH}_3\text{CH}_2\text{CH}-\text{CH}-\text{COOH}}{\overset{\text{CH}_3 \quad \text{NH}_2}{}}$$
Serine	Ser	$$\underset{\text{CH}_2-\text{CH}-\text{COOH}}{\overset{\text{OH} \quad \text{NH}_2}{}}$$
Threonine	Thr	$$\underset{\text{CH}_3\text{CH}-\text{CH}-\text{COOH}}{\overset{\text{OH} \quad \text{NH}_2}{}}$$
Aspartic acid	Asp	$$\underset{\text{CH}_2 - \text{CH}-\text{COOH}}{\overset{\text{COOH} \quad \text{NH}_2}{}}$$
Asparagine	Asn	$$\underset{\text{CH}_2 - \text{CH}-\text{COOH}}{\overset{\text{CONH}_2 \quad \text{NH}_2}{}}$$
Glutamic acid	Glu	$$\underset{\text{CH}_2\text{CH}_2-\text{CH}-\text{COOH}}{\overset{\text{COOH} \qquad \text{NH}_2}{}}$$
Glutamine	Gln	$$\underset{\text{CH}_2\text{CH}_2-\text{CH}-\text{COOH}}{\overset{\text{CONH}_2 \qquad \text{NH}_2}{}}$$
Lysine	Lys	$$\underset{\text{CH}_2\text{CH}_2\text{CH}_2\text{CH}_2-\text{CH}-\text{COOH}}{\overset{\text{NH}_2 \qquad\qquad \text{NH}_2}{}}$$
Hydroxylysine	—	$$\underset{\text{CH}_2-\text{CH CH}_2\text{CH}_2-\text{CH}-\text{COOH}}{\overset{\text{NH}_2 \text{ OH} \qquad\qquad \text{NH}_2}{}}$$
Arginine	Arg	$$\underset{\underset{\text{NH}}{\overset{\|}{}}}{\overset{\text{NH}_2 \qquad\qquad\quad \text{NH}_2}{\text{CNHCH}_2\text{CH}_2\text{CH}_2-\text{CH}-\text{COOH}}}$$

Fig. 7.2 (*cont.*). Some commonly occurring amino-acids

Trivial name	Shortened name	Formula
Cysteine	CySH	$\overset{\displaystyle SH}{\underset{\displaystyle |}{}}\ \overset{\displaystyle NH_2}{\underset{\displaystyle |}{}}$ $CH_2-CH-COOH$
Cystine	CySSCy	$\overset{\displaystyle NH_2}{\underset{\displaystyle |}{}}$ $S-CH_2-CH-COOH$ $\underset{\displaystyle |}{}$ $S-CH_2-CH-COOH$ $\underset{\displaystyle NH_2}{\underset{\displaystyle |}{}}$
Methionine	Met	$\overset{\displaystyle S-CH_3}{\underset{\displaystyle |}{}}\quad \overset{\displaystyle NH_2}{\underset{\displaystyle |}{}}$ $CH_2CH_2-CH-COOH$
Phenyl-alanine	Phe	(benzene ring)$-CH_2-\overset{\displaystyle NH_2}{\underset{\displaystyle |}{}}CH-COOH$
Tyrosine	Tyr	$HO-$(benzene ring)$-CH_2-\overset{\displaystyle NH_2}{\underset{\displaystyle |}{}}CH-COOH$
Tryptophan	Try	(indole ring) $C-CH_2-\overset{\displaystyle NH_2}{\underset{\displaystyle |}{}}CH-COOH$ $\underset{\displaystyle CH}{}$ N H
Histidine	His	$N===CH$ $HC\quad\quad C-CH_2-\overset{\displaystyle NH_2}{\underset{\displaystyle |}{}}CH-COOH$ N H
Proline	Pro	$\overset{\displaystyle CH_2}{}$ $H_2C\quad NH$ $H_2C\quad CH-COOH$ CH_2
Hydroxy-proline	Hypro	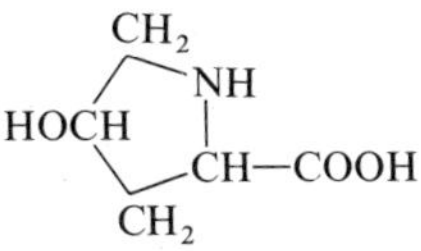 $\overset{\displaystyle CH_2}{}$ $HOCH\quad NH$ $\quad\quad CH-COOH$ CH_2

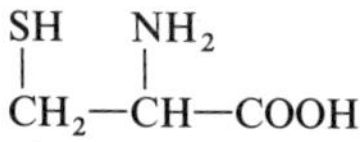

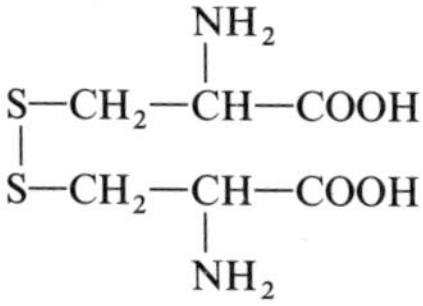

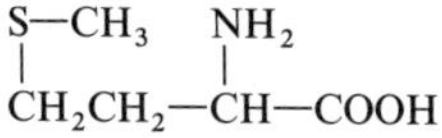

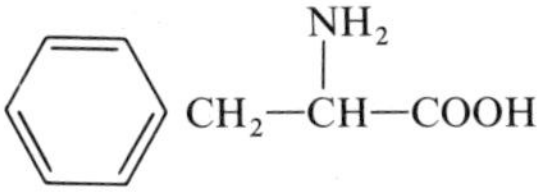

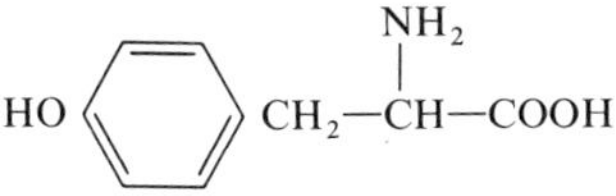

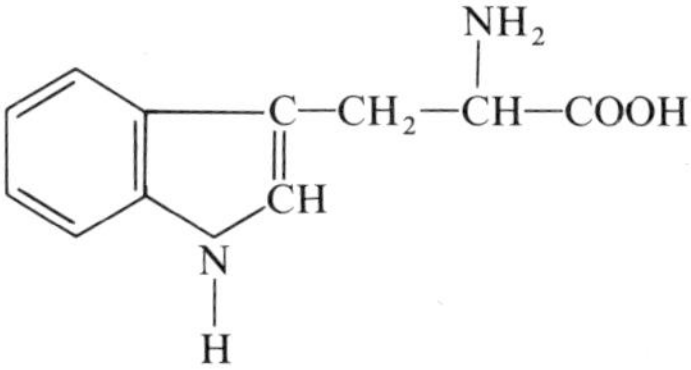

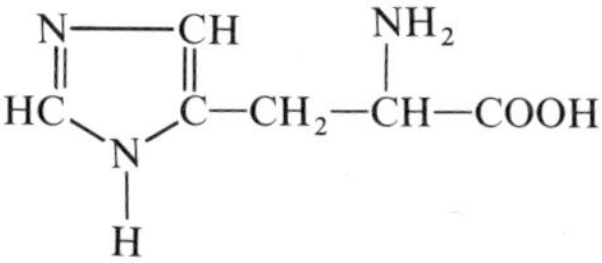

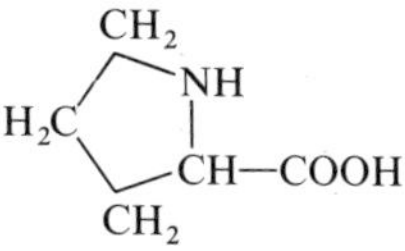

Fig. 7.2 (*cont.*).

Fig. 7.2 (*cont.*) The three-dimensional arrangement of amino-acids in a peptide chain. R₁, R₂, R₃ are groups containing C, H. The bond lengths are in Å

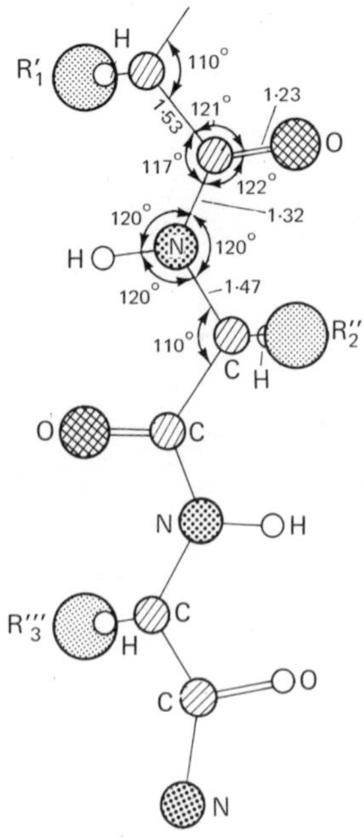

valine, leucine and isoleucine, the length of the hydrocarbon side chain increases. So far the side chain only contains carbon and hydrogen atoms and is essentially non-polar. The side chain protrudes from the polypeptide chain and may be regarded as being used to separate the folds in the polypeptide chain. In cystine(cys) it will be noted that there are two sulphur atoms which are linked together.

The structure of proteins may be looked at from four different stages. First the primary structure is concerned with the order of the amino-acids in the polypeptide chain, and secondly the secondary structure is concerned with the internal structure of the polypeptide chain. The tertiary structure is concerned with the arrangements of the side chains and it is often found that the non-polar parts point inside while the polar parts point outside. The quaternary structure of the proteins is concerned with how the aggregations of the molecules are related and we shall see that there are certain simple elements of symmetry concerning these. Some of the questions we must ask ourselves are: Why is it important to know the structure of these proteins? Does X-ray analysis give us the complete answer to the structure? Is there any order in the structure and indeed how do you study these very complex structures? A knowledge of the structure of proteins and related compounds, apart from its purely academic interest, must lead us to a greater understanding of certain diseases. Insulin is necessary for the control of diabetes. Haemoglobin and myoglobin are directly concerned with the uptake of oxygen and an understanding of their structure might help in the treatment of certain respiratory diseases. The essential component of genes is not made of protein but of nucleic acids. Nucleic acids have two forms, DNA which is found in genes and in some viruses such as those of smallpox and chickenpox, and RNA found in the ribosomes on which new protein molecules are built in the cell and other viruses such as those of influenza and polio. The interaction of nucleic acids and proteins, however, directly influences biological replication and metabolic activity and an understanding of these two and the relationship between them is of vital importance. Pernicious anaemia was a very worrying disease at the turn of the century and it was not until vitamin B_{12} was isolated and studied that the treatment of this illness was successful. The last of the three questions is answered in the following sections.

The primary structure (insulin)

The primary structure of the polypeptide chain involves the arrangement of the amino-acids and the interlinking between the polypeptide chains by disulphide linkages. As such the problem is far too difficult to be studied successfully at the moment by X-ray analysis. Some indication of the possible structure is given by the study of simple peptides and it is possible that a knowledge of these structures (Fig. 7.2) will be of great help in understanding the full structure of the polypeptide chain. The carbon and nitrogen main skeletal chain is not linear because the alpha carbon is tetrahedral, and is surrounded as symmetrically as possible by the four groups. The carbon of the carboxyl group is surrounded in a plane by three atoms, i.e. the alpha carbon, the oxygen and a nitrogen but the atoms are not at the corners of a regular triangle as the bond lengths of the carbon to the three respective atoms are different. The nitrogen is surrounded by three groups in a plane, similarly, and the bond angles between H–N–C, etc., are 120°. The study of proteins has been enhanced by the development of chromatographic methods which were used by Sanger in the study of insulin. Sanger was able to elucidate completely the sequence of the fifty-one residues of amino-acids which are in two S—S bonded chains. The residues are arranged in a definite genetically determined order but the sequence itself is not simple. Between the two chains there are cystine bridges (Fig. 7.3) in which the linking unit is —CH_2—S—S—CH_2— and is often referred to as the disulphide linkage, an important feature of which is that the bonds are covalent with the bond energies of 400 kJ mol^{-1} but which can be hydrolysed easily. Later on in the structure of DNA we shall find that the main bonds between the helical nucleotide chains are the much weaker hydrogen bonds with an energy of 20–40 kJ mol^{-1}. In insulin the disulphide bridges are strong but they can be easily broken by hydrolysis. Sanger extended his brilliant chemical analysis to even more complex molecules than insulin and the amino-acid

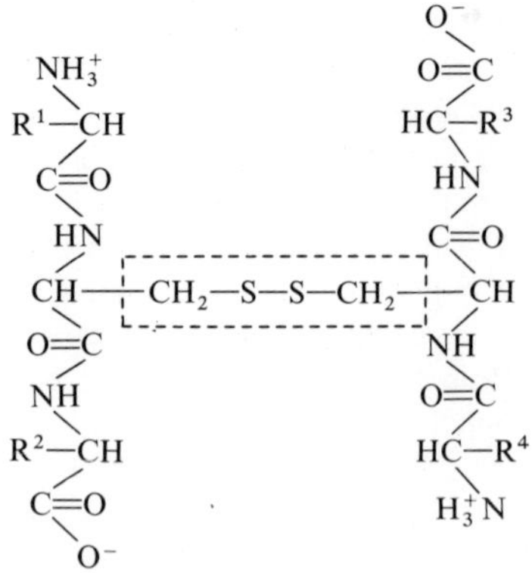

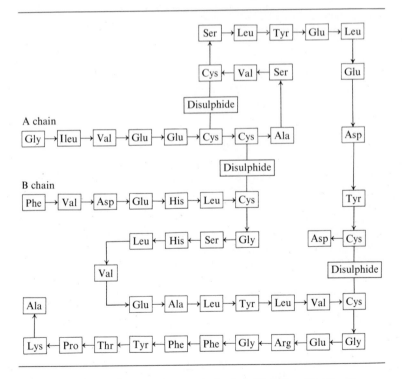

Fig. 7.3. A typical disulphide bridge between peptides

Fig. 7.4. The amino-acid sequences in the A and B chains of beef insulin
The disulphide bonds are common linkages between chains. The bond indicated by
an arrow is a peptide bond between CO and NH groups

sequences in the tobacco mosaic virus ribonuclease and chymotrypsin have now been determined.

But what exactly is insulin? The hormone insulin may be isolated from the pancreas and is the hormone which is vital to control the glucose metabolism and to relieve the symptoms of patients suffering from diabetes. The molecular weight is very high although small for proteins being approximately six thousand which only varies a little when obtained from different species. An understanding of the X-ray analysis is helped by the previous knowledge of the primary structure of insulin (Fig. 7.4). One of the main tasks, however, in the chemical study of insulin is to relate details of the specific reactions which are controlled by the hormone to the structure. The hormone does influence some metabolic processes, protein and nucleic acid syntheses, and nucleide synthese and the uptake and metabolism of glucose by cells, in all of which it is probable that the hormone bonds directly with the glucose. An X-ray study will lead to an understanding of how a large molecule like insulin reacts with a small molecule like glucose.

Why is it not possible to write down and draw the full crystal structure since the primary structure is known? The main reasons are that the polypeptide chains are not packed symmetrically. Whereas X-ray analysis is greatly helped by the complementary chemical knowledge it is still necessary to carry out the full analysis which involves preparing a good single crystal; taking many reflections along the main axes; measuring the position and intensity of the diffraction spots; using Fourier analysis together with reliable chemical knowledge to calculate the coordinates of the atoms in the unit cell whose dimensions must be determined.

The X-ray analysis of insulin has not been completed but the following illustrates the type of problem met. The main problem concerns the intensity of the spots, which we will consider again in connection with haemoglobin. When insulin is crystallised in the presence of zinc ions the zinc is incorporated into the crystal in a random manner with one or two ions per unit cell. The intensity of the diffraction spots is influenced by the ions and using Fourier methods attempts are made to calculate the electron density within the crystal. The trigonal unit cell contains six molecules of insulin and the zinc ions may be removed by a chemical called EDT (ethylene diamine tetra-acetic acid). The intensity of the spots from the two crystals are compared leading to a value of the phase angle which was discussed in Chapter 3.

Resolution. What is meant by the term resolution? Resolution means to what extent the atoms or groups of atoms may be seen or not seen as individual units. So far with insulin a resolution of 0·45 nm (4·5 Å) has been obtained, but since the average bond length is much smaller than this 0·5 nm (1·5 Å) this resolution does not distinguish between the atoms. The average distance between the polypeptide chains, however, is more than 0·6 nm (6 Å) so the resolution can tell the crystallographer where the polypeptide chains are in the unit cell.

178

The secondary structure (α-helix)

The secondary structure of proteins is concerned with the configuration of the polypeptide chain. No configuration is stable unless the configuration allows the carbonyl group (which belongs to the same chain as the NH group or a different chain) to be hydrogen bonded to the imino group. The

Fig. 7.5. Some hydrogen bonds in keratin
The structure is basically a helix. The hydrogen bonds are shown as the interlinking strands

structure of many long chain polymers which have repeating atomic arrangements often have a screw symmetry or a helical configuration. When hydrogen bonding may occur between the groups in the same chain all structures are helical. Pauling, Corey and Bransom discovered the alpha helix, in which there is a right-handed screw axis, to be generally more stable than the helix in which there is a left-handed screw axis. Keratin has a right-handed alpha helical structure (Fig. 7.5) with hydrogen bonds as interlinking strands and as keratin contains a high proportion of cystine there are additional disulphide bridges between the coils of the helix. In this structure each N—H is hydrogen bonded to an O=C group which is three amino-acid units along the chain and the bond is parallel to the helical axis. There are five complete turns of the helix for each eighteen of the amino-acid residues and X-ray diffraction photographs contain a spacing between rows of spots—attributed to a distance of 0·15 nm (1·5 Å)—which is the repeating distance of the amino-acid residues along the axis of the helix. The pitch of the helix has a spacing of 0·54 nm (5·4 Å) but this is found to be slightly less in keratin which is wound up like a coiled spring. The coiled form of keratin is called alpha-keratin and occurs in hair, horn and nails but it is stretched beta-keratin which occurs in silk fibroin.

In a typical alpha-helix (Plates 1.4 and 7.1) there are more intermolecular hydrogen bonds than the form with the lefthanded spiral. In the plate

Plate 7.1. The alpha-helix

the black balls represent the carbon atoms and following these up from the bottom right will give a spiral.

Hydrogen bonds are very important. They are weak bonds (40 kJ mol^{-1}) and the bond distance between neighbouring atoms in the hydrogen bond is of the order of 0·3 nm (3 Å). When compared with a value of 400 kJ mol^{-1} and a bond length of 0·15 nm (1·5 Å) in the covalent bond it is realised that the hydrogen bonds are weak. Nevertheless hydrogen bonding governs the struc-

180

ture of molecules such as nylon, DNA and proteins, and an understanding of the mechanism by which it is formed is very important. The hydrogen bond is formed when a hydrogen atom is joined to an electronegative atom such as nitrogen or oxygen. (Electronegativity is defined as the ability of an atom in a bond to attract electrons to itself.) The electronegative element therefore attracts more electrons than does the hydrogen which is left with a partial positive charge. At the other end of the hydrogen bond is an electronegative group which has a partial negative charge and the hydrogen bond is formed by the attraction between the partial negative and the partial positive charge. For example, in acidified water (Fig. 7.6) each oxygen has a partial negative charge and each hydrogen carries a partial positive charge. The result is that in ice crystals, and to some extent in water, molecules aggregate together, giving a structure which has hexagonal symmetry.

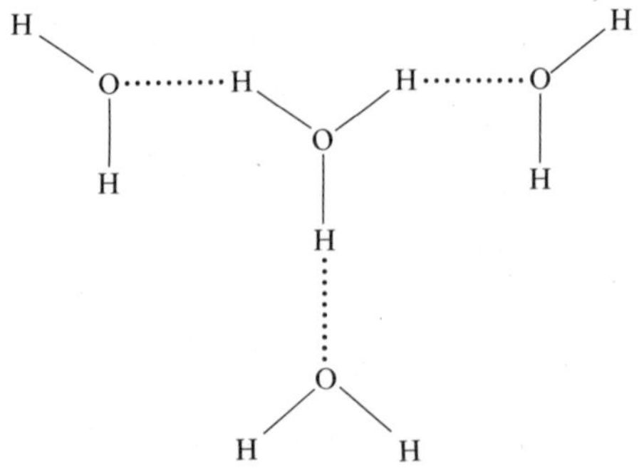

The HOH angle in the H_3O^+ ion is about 115° and each ion binds three more water molecules by forming hydrogen bonds. The aqueous proton therefore has an average composition of $[H_9O_4]^+$ which has a lifetime of about 10^{-13} seconds.

Fig. 7.6. Hydrogen bonds in acidified water

In the protein molecules the hydrogen bond is formed between the imino group and the carbonyl group, the distance between the imino-hydrogen and the carbonyl-oxygen is of the order of 0·3 nm (3 Å). Often one finds that the polar surfaces which are capable of forming hydrogen bonds with solvents like water are turned towards the outside of the molecule, whereas the non-polar parts are directed toward the inside of the molecule where few, if any, solvent molecules are present. Hydrogen bonding forms the major part of the governing factors in replication of protein molecules which is governed by DNA. Hydrogen bonding occurs between the chains of polyamides in nylon, a polymer which forms fibres that are used as substitutes for natural fibres, e.g. ropes, and are suitable for clothing. The molecules of the fibre are highly orientated and because of this the nylon is fibrous. As soon as the symmetry of the molecules is lost so are the fibrous properties. The fibre forming properties of nylon result in the main from the interchain hydrogen bonding which results in the molecular linearity (Fig. 7.7, see also 7.18 and 7.19). If the main skeletal $(CH_2)_n$ chain of the nylon is increased, that is the number of hydrogen bonds is reduced, then the melting point lowers but the material remains a fibre. If, however, side chains are introduced which has the effect of pushing the chains apart, then as the degree of alkylation or the length of the side chain is increased, the nylon type polymer no longer retains its fibre forming properties and becomes a rubber in which the molecules are intertwined. When all the amide groups are substituted nylon becomes a viscous liquid.

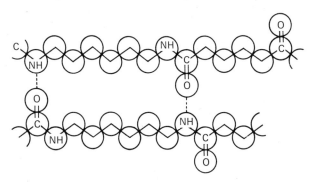

Fig. 7.7. Nylon showing intermolecular hydrogen bonding

One of the main problems of the crystallographer is to distinguish between a straight chain, as in nylon, or an alpha-helix as in keratin. Luckily the presence of an alpha-helix in a fibrous material may be recognised quite easily by X-ray analysis. The repeat distance of the amino-acid residues occurs, as we have seen in keratin at 0·15 nm (1·5 Å) when measured along the axis and gives rise to strong X-ray reflections from planes which are perpendicular to the fibre axis. The right-handed alpha helix, therefore, may be considered rather like a spiral staircase in which the amino-acid residues form the individual steps. The height of each turn is 0·54 nm (5·4 Å) high. It takes eighteen steps to return to the position of the same symmetry as the point from which one started. The helix is mainly held together by hydrogen bonds between the carbonyl groups of one residue and the imino group of the fourth residue along the chain.

The tertiary structure (haemoglobin, myoglobin and lysozyme)

The polypeptide chains in spite of their secondary structure are often folded into nearly spherical molecules. It is possible to obtain X-ray diffraction photographs of crystalline enzymes indicating that the structure of the enzymes is well organised and most of the thousands of atoms which constitute an enzyme molecule are fitted into a definite place. The main problem of crystal X-ray analysis is to determine the positions of these atoms but the situation is rendered even more complex when the molecule contains many thousands of atoms.

Haemoglobin. Haemoglobin is a protein of molecular weight 64 500 with four iron atoms in ten thousand atoms. These iron atoms are combined with protoporphyrin to give four haem groups and the remaining atoms are in four polypeptide chains which each contain approximately 140 amino-acid residues.

Such is the conclusion of many years work by Perutz but the main problem here is to see, non-mathematically, how such results were achieved. In 1937 Bernal and Hodgkin showed that crystals of proteins could be grown which diffracted X-rays like simpler crystalline substances. It was considered possible that the arrangements of the atoms might be determined but the problem was to interpret the diffraction patterns. Sir Lawrence Bragg supported Perutz and Kendrew when they decided to study haemoglobin and myoglobin respectively.

Perutz mounted a crystal of haemoglobin in a glass capillary to keep it wet, and an X-ray diffraction photograph was obtained by rotating the crystal in the beam in the usual manner. The spots lie at the corners of a regular lattice which bears a reciprocal relationship to the atomic spacing between the planes of the atoms. The closer the atoms along the axis being investigated the wider the spacing of the diffraction spots and it is possible to calculate the positions from the spacing of the spots provided the intensity of the spots and the phase problem for the image are known. The arrangement of the atoms in the unit cell determines the strength of certain reflections, because the diffracted waves of X-rays may reinforce or cancel each other out.

In Fig. 7.8 there are four fringes which have the same wavelength and amplitude but which differ in four different values of the phase angle. The problem of combining such waves is solved in two main stages. First the set of fringes is correctly placed with respect to some common origin but at this origin the wave may be a maximum, a minimum or at some intermediate stage of the phase angle. Great care must be taken at this stage because with a given set of fringes of fixed amplitude an infinite number of arrangements of the atoms may be deduced. The next stage is to fix the phase angle relationship at the origin which is usually attempted using the method of isomorphous substitution.

The presence of a heavy metal atom greatly influences the intensity of the diffraction spots because the X-rays interfere with the electrons outside the nucleus. Atoms such as hydrogen which only has one electron have a much reduced influence on the intensity of the spots, and it is often very difficult to determine the positions of hydrogen atoms. A crystal of the substance is therefore grown so that a heavy metal atom is randomly included into the crystal as happened when zinc was included into the crystal of insulin. X-ray patterns therefore can be compared to those from the original crystal when a series of heavy metal atoms is present in separate crystals so that certains spots appear to have different intensities. It is possible to calculate the distance of the wave maximum from the heavy metal atom and then determine the value of the phase angle but this leaves open the question of whether there is a maximum or a minimum at the origin. A second crystal with a heavy metal atom enclosed in the unit cell is grown and in certain crystals it is possible to determine whether there is a maximum or a minimum.

Fig. 7.8. Identical fringes with four different values of the phase angle

Having determined the phase angle the next stage is to relate this pattern to the crystal structure and it is important to realise that the diffraction pattern is rather like a slice through a fruit cake. The slice is in a plane that is in two dimensions whereas the cake occupies a volume and is in three dimensions. The crystallographer attempts to relate the two dimensional pattern to the three dimensional cell. A large number of photographs are taken to calculate the electron density of the atoms in the unit cell and the results are often plotted manually on sheets of perspex as lines which represent the regions of the same electron density. A heavy metal atom is present when the lines or contours are close together and a less heavy atom when the rings are widely spaced. The polypeptide chains may be traced throughout the unit cell because the rings of electron density move steadily through the electron density map, but hydrogen atoms are difficult to see because of their low electron density. The position of the side chains may be seen when the resolution is 0·2 nm (2 Å) and in the study of myoglobin Kendrew has been able to determine which side chains R are present by looking at the shapes of the electron density distribution. A methyl group

184

has a different electron density distribution from an isopropyl group because the latter group has more carbon atoms. These results compare very favourably with a chromatographic separation of the amino-acid residues, because computers aided the calculations. The development of computers has facilitated the X-ray analysis of many structures but one important problem is the storing of the great wealth of information. With haemoglobin many spots have to be studied. The intensity is measured and with myoglobin at 0·2 nm (2 Å) resolution one quarter of a million spots were studied, from which the amplitudes were calculated, phase angles and signs (maximum or minimum) determined and coordinates calculated.

Plate 7.2. Haemoglobin—Two pairs of polypeptide chains related by the diad axis
The arrow shows how one pair is placed over the other to assemble the complete molecule

The function of haemoglobin is to take oxygen from the lungs to the tissues and return carbon dioxide to the lungs, there is one molecule of myoglobin to one molecule of oxygen. The ferrous ions in haemoglobin combine reversibly with the oxygen but are not oxidised themselves because the oxygen remains as molecular oxygen. The ferrous ions are not isolated as the oxygen uptake with four free iron units is not four times as powerful as with one free ion.

The electron density model tells us that the molecule consists of four polypeptide chains of roughly equal length and there are two identical pairs with each chain symmetrically related to one partner. The first stage in the formation of haemoglobin is in the matching of the pairs (Plate 7.2) and then the pair of chains which are illustrated white is placed over the black pair resulting in an overall symmetry of a tetrahedron. It is interesting to note that although the chains themselves are irregularly arranged they do form a symmetrical unit, and that as a consequence of this order the ferrous ions are widely spaced. There is one twofold axis of symmetry and a hole runs right through this axis, but the haem groups lie in four separate spaces (Fig. 7.9) at the surface of the molecule while the closest distance between the atoms of iron is 2·5 nm (25 Å) which is much larger than expected. Since the haem groups are easily reached by oxygen in the transport of oxygen it is not surprising to find them at the surface. The model at first sight however does not explain how the uptake of oxygen in one atom is not completely independent of the uptake on another. Usually one finds that interactions only occur through two or three atoms that is through 0·3 nm (3 Å) but certainly not over the large distance as in haemoglobin. How does the structure explain the uptake of oxygen? The answer is that when haemoglobin takes up oxygen to form oxyhaemoglobin the arrangement of the peptide chains changes and thus affects the structure of the molecule.

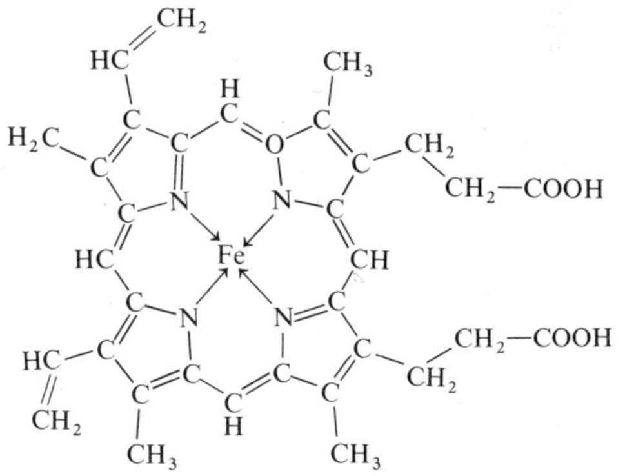

Fig. 7.9. The structure of the haem group

In the haem group the ferrous ion is surrounded by four nitrogen atoms which are in a plane, but the iron is probably 0·02 nm (0·2 Å) out of the plane. The fifth position around the iron is occupied by a nitrogenous group of the globin called a histidine residue and the sixth group is a water molecule which may be replaced by a molecule of oxygen. How delicate is this uptake of oxygen? The reversible uptake is extremely finely balanced and a small change in the structure has a very large effect. If the ferrous ion is changed for a ferric ion then no oxygen is taken up. Does the carbon dioxide molecule directly replace the oxygen molecule? The evidence available suggests that the ferrous ions do not carry the carbon dioxide directly, but it is not fully understood how the molecule is transported.

Myoglobin. Myoglobin is the substance which colours vertebrate muscle red and transports oxygen like haemoglobin; it is regarded as an oxygen-storer rather than an enzyme although the distinction is not clear. Kendrew chose the myoglobin from the sperm whale as whale meat is one of the richest sources of myoglobin. Myoglobin has a molecular weight of approximately 17 000 and consists of a globin which is a single polypeptide chain with 153 residues and one haem group which consists of a protoporphyrin

186

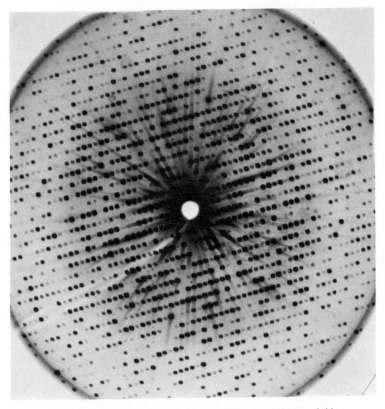

Plate 7.3. X-ray diffraction photograph of sperm whale myoglobin

which is a substituted pyrolic ring. In addition there is an iron unit which is attached to five of the neighbouring nitrogen atoms. Four of these nitrogen atoms are attached to the pyrolic rings and the other is in the histidine. The remaining group which is attached to the iron is the water molecule which may be replaced by the molecular oxygen.

What factors govern the X-ray crystallographers' choice of the crystal? The main requirement is that the protein should be of as low a molecular weight as possible, and easily prepared and crystallised, and sperm whale myoglobin filled most of these requirements. How does myoglobin resemble haemoglobin? Myoglobin consists of 150 amino-acids in a single poly-peptide chain and four similar units constitute the structure of haemoglobin whose molecular weight is four times that of myoglobin. With four ferrous ions in haemoglobin but only one in myoglobin, one can see that there is a one to four relationship between myoglobin and haemoglobin.

At the start of the structural determination two-dimensional represen-tations were attempted but although easier than three dimensional studies they did not yield much information because the electron density map showed too much overlapping of one atom over another. The three-dimensional study involved taking a large number of photographs and calculations of phases and the final 0·15 nm (1·5 Å) resolution will involve taking 25 000 reflections. When the work was begun however, no computer was fast enough to handle the calculation and a resolution of 0·6 nm (6 Å) was decided upon which involved 400 reflections. Such a resolution as we have seen illustrates the position of the polypeptide chain and the haem group around which there are regions of high electron density. The diffraction pattern (Plate 7.3) consists of rows of spots of differing intensity and it is from these spots that the structure is calculated as each photograph corresponds to a two dimensional reflection through the three-dimensional lattice and corresponds to a single Fourier component. The complete structure of myoglobin is evaluated using all the components in the Fourier synthesis but the spots of higher order which fill in the fine structure lie at the outside of the pattern and are not considered in the first low resolution analysis. The low resolution analysis considers the spots within a circle at the centre, and the results of the Fourier analysis are plotted on the electron density map (Plate 7.4).

The polypeptide chains appear as rodlike segments which are joined at the ends and the single dense flattened disc is the porphyrin ring which surrounds the ferrous ion. In between the regions of high electron density there is liquid, but the chain may be followed as a continuous region of high electron density which is irregularly distributed and totally lacking in symmetry. This irregularity compares with haemoglobin in which the four sub-units closely resemble the myoglobin molecule.

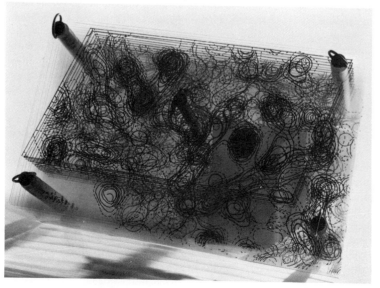

Plate 7.4. Electron density map of myoglobin at 6 Å resolution

At higher resolution the polypeptide chains appear as hollow cylinders with the alpha-helix visible showing that myoglobin belongs to the family of alpha-fibrous proteins. The helical segments are mostly precisely straight and it is even possible, as indicated previously, to see the orientation of the side-chains relative to the helical axis. From a knowledge of the configuration of the L-amino-acid it is possible to show that the helix is indeed right handed. The iron unit is 0·02 nm (0·2 Å) out of the plane of the porphyrin group and histidine is the haem-linked group on the opposite side to the molecular oxygen. Three quarters of the atoms have been assigned special coordinates which were obtained by the method of approximations. In this method the coordinates of the atoms are approximately assigned from which the phases of the reflections can be calculated which in turn are combined with the observed amplitudes to give the first refinement. In each refinement the coordinates of more atoms are assigned.

How does myoglobin differ from haemoglobin? Apart from the one to four ratio that has been mentioned myoglobin is a compact molecule and there are no water molecules inside apart from a few which are caught as the molecule folds up. There are no channels through the molecule, and the volume of the internal free space is very small with the non-polar side chains turned inside but polar point outside. The whole structure is held together by the van der Waal bonds between the non-polar chains.

Lysozyme. Lysozyme is the first enzyme whose three-dimensional structure has been evaluated. But let us pause for a while to learn how Alexander Fleming isolated this enzyme. In 1922, seven years before he demonstrated the existence of penicillin, he was attempting to find the universal antibiotic. When he was suffering from a cold he placed some nasal mucus onto a culture of bacteria and was delighted to find that the bacteria near the mucus had dissolved away. After working very hard he managed to demonstrate that the antibacterial action was due to an enzyme which he called lysozyme as it was able to lyse or dissolve the bacteria. Lysozyme was not however the universal antibiotic that he had hoped for because it is unable to attack the most harmful bacteria.

Like all enzymes we now know that lysozyme is a protein—here there is but one chain of 129 residues with twenty different amino-acids and there are four disulphide bridges between cystine residues. In 1960 an Argentine chemist who was working at the Royal Institution in London showed that suitable crystals containing heavy metal atoms could be prepared so that in principle it was thought that it would be possible to solve the phase problem of the X-ray crystallographers. The first low resolution image of the electron density which took two years to complete, showed that the general arrangement of the polypeptide chain is even more complex than that of haemoglobin and myoglobin. This low resolution study was carried out at the Royal Institution by D. C. Phillips and co-workers, and the study involved calculations from the amplitudes of about 400 diffraction maxima measured from the native protein and from three further types of crystals, which each

189

Plate 7.5. Lysozyme

contained a different heavy metal atom. The next stage, which was completed in 1965, gave an image calculated from 10 000 diffraction maxima which resolved features separated by 0·2 nm (2 Å). The advantage of such a resolution is that it is sufficiently high for the positions of many groups to be recognised by the shapes of the calculated electron density patterns. The disadvantage however is that the positions of individual atoms in the polypeptide chains cannot be identified and except for the trapped chloride ions the position of all atoms was inferred from the electron density image, together with the available chemical knowledge. The chloride ions are trapped by the enzyme as it forms in the solution.

Are there any recognisable features of the structure and does this structure help to explain the way in which lysozyme destroys the cell walls of the bacteria? The first fifty-six residues in the chain are more symmetrically arranged than the remaining ones and we will tend to limit our discussion to

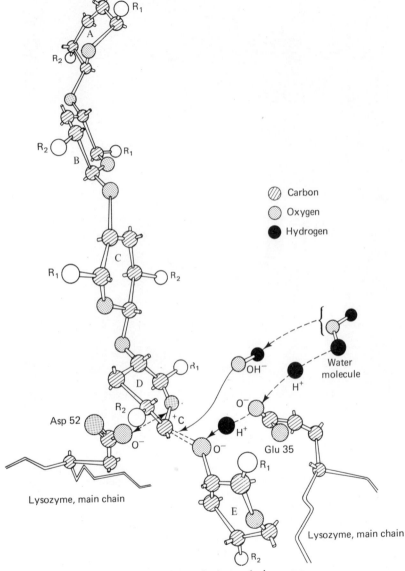

Fig. 7.10. Splitting of substrate by lysozyme

these. The observed conformation is formed such that the polar parts (NH) groups of the molecule point towards the solvent or surrounding liquid. The non-polar (CH_2) groups are kept away from the surrounding liquid. Hydrogen bonds between residues in the same chain play a profoundly important part in determining the arrangement of the chain. Thus the first forty residues are kept in a fairly tight structure because of the hydrogen bond between residues 1 and 40. The chain is asymmetrically distributed, and the overall structure is that there is a large fold or cleft in the middle of the molecule and it is in this fold that the sugar residues from the bacterial cell walls fit.

It is known that lysozyme attacks the sugar residue in the cell wall and Louise Johnson, again at the Royal Institution, managed to prepare some crystals in which sugar molecules were bonded to the lysozyme. She added sugar to the mother liquor from which crystals of the enzyme had been prepared and in which they were suspended. The sugar molecules diffuse into the protein crystals along the fold of the molecule in which there are channels which run right through the crystal and which are normally filled with water. The sugar molecules fit into the fold without altering the overall structure of the lysozyme crystal. Studies of the X-ray diffraction patterns at low resolution and at 0·2 nm (2 Å) resolution have indicated how the sugar is bonded to the enzyme. Useful images of the electron density have been calculated from the changes in the amplitude of the diffracted waves assuming that the phase relationships are the same in the pure protein and with the protein containing the sugar molecule. Unfortunately X-ray studies cannot indicate the mechanism of a chemical reaction but it is possible to guess at the mechanism. What probably happens is that the glutamic acid (residue 35) releases a proton (H^+) which protonates the oxygen atom between the two sugar rings. The bond between the two rings is broken leaving a carbonium ion (C^+) on the ring attached to the sugar molecule but the rings do not finally move apart until the glutamic acid residue regains a proton. Probably the carbonium ion is stabilised by the negative side chain of aspartic acid (residue 52) but then drags or polarises an OH^- group from water leaving the proton free to go to the residue 35 so that the rings may move apart. The cycle restarts when the sugar molecule is in the correct position after it has diffused further through the fold of the enzyme. For further study read 'Lysozyme' by D. C. Phillips, *Scientific American*, **78**, 1966, 215.

The quaternary structure of proteins (enzymes)

The quaternary structure results from the interaction between two or more polypeptide chains, and as was seen in haemoglobin, the structures are often symmetrical. Identical sub-units (Fig. 7.11) may be linked together by mutually reciprocal regions and the result is a structure which always has the symmetrical twofold, threefold or fourfold cyclic symmetry. The electron microscope is particularly useful in studying the quaternary structure and in typical results (Plates 7.6 and 7.7) the liver enzyme has the fourfold cyclic symmetry with four sub-units per molecule as there is only one layer. In some structures, however, there are more than one layer and in the enzyme glutamine synthetase from the gut bacillus *E. coli* there are twelve sub-units which are arranged in hexagonal units in two layers. Two perpendicular views are shown, the lower one looks down on the molecule along a sixfold rotation axis and shows the arrangement of the six sub-units in one

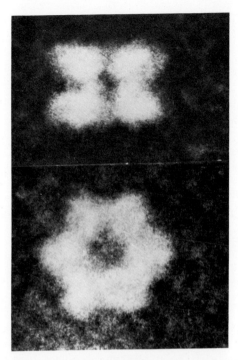

Plate 7.6. The molecule of an enzyme (glutamine synthetase) from the gut bacillus *E. coli.* Twelve sub-units arranged in two hexagonal layers. Two perpendicular views are shown. Magnification × 3 000 000

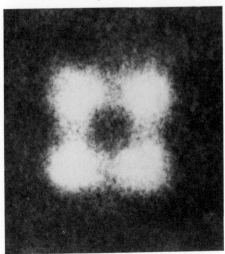

Plate 7.7. The molecule of an enzyme (pyruvate carboxylase) obtained from liver. Four sub-units arranged at the corners of a square. Magnification × 3 000 000

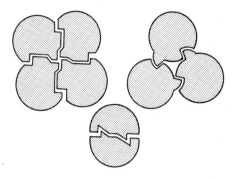

Fig. 7.11. The quaternary structure of proteins.
Fourfold, threefold, twofold cyclic rotational symmetry

layer, the upper one looks along one of the twofold axes through the two layers. Plate 7.8 shows catalase molecules formed into a regular array which will go on to produce a crystal. Ultimately the crystal may grow large enough to be seen with a magnifying glass. Catalase is an enzyme, one of the many hundreds now known. Each is a protein which acts as a catalyst specific for one particular biochemical reaction. This packing of layers may lead to a helical structure as is shown in Plate 7.8 for catalase, so here at least there is some optical evidence for the helical structures which X-ray diffraction has very strongly indicated as being common among many proteins. There is powerful indication that the quaternary structure is vital to the enzymes and their catalytic activity. Mild treatment, for example in heating or changing pH will often dissociate the enzymes into their individual sub-units and the enzymes are then found to be inactive. When the process is reversed and the sub-units are rejoined together under favourable conditions then the enzyme activity returns because the sub-units rearrange themselves into their original symmetrical aggregates.

The structure and replication of deoxyribonucleic acid (DNA)

Genetic information is carried by the chromosomes which consist chemically of DNA and protein. In 1962 Watson, Crick and Wilkins shared the Nobel Prize for Medicine and Physiology for their work in determining the structure of DNA. Watson in his book *The Double Helix* has described how he persuaded the finest partners he could find to join him in resolving the structure of DNA. He had been interested in the problem of the chemical structure of genes for a long time but it was not until Wilkins described to him how an X-ray diffraction pattern could be obtained from DNA that it was realised that DNA was a crystalline ordered substance. As we have seen, Pauling partly solved the structure of proteins in terms of the alpha helix

194

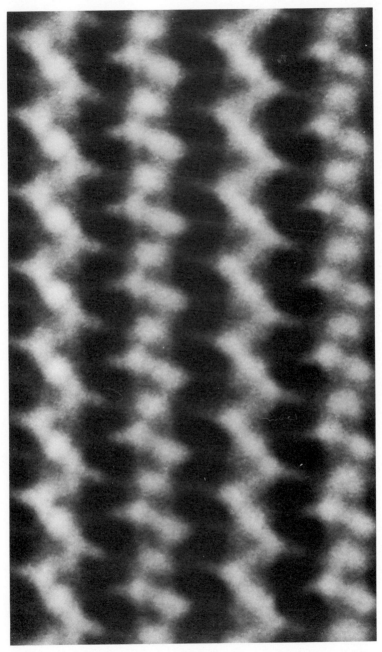

Plate 7.8. The regular arrangement of protein units seen in a crystal of catalase magnified more than a million times with an electron microscope

and this was a major step in solving the structure of DNA. Crick and Watson set themselves a task of constructing a model which would explain the diffraction patterns of DNA which were obtained by Wilkins. At certain times great problems were met because the work was not getting very far, and Crick was studying for his PhD on a different problem. Eventually it became clear that the structure was not a single helix but a double helix with the two chains of the DNA molecule spiralling in opposite senses. A note was then published in *Nature* and the article ended in the rather dubious sentence, 'It has not escaped our notice that the specific pairing we have postulated immediately suggests a possible copying mechanism for the genetic material'. This means that it was realised that the double helix somehow explained how the genetic information could be carried by the DNA in the chromosomes.

Chemically DNA is a long-chain polymer with a skeleton which repeats at regular intervals. The chain consists of phosphate groups which join successive deoxyribose residues, each of which is attached to one of four different kinds of nitrogenous bases. There are two purines, Adenine (A) and Guanine (G), and the pyrimidines, thymine (T) and cytosine (C). It is only the bases which are variable constituents in the skeletal chain and in some way the bases or their sequence must carry the genetic information. Although the sequence of bases AGCT is not known, it is known that the ratio A:T and G:C = 1:1 irrespective of the source of the DNA.

The structure of DNA (Plate 7.1 and Plate 1.4) then consists of two chains that run in opposite directions and are coiled around each other to form a double helix. The bases with their large hydrophobic (or water hating) surfaces are neatly stacked together on the inside, with the phosphate residues facing the aqueous medium. Each step of the staircase contains two bases, one from each of the chains, which are linked together by hydrogen bonds, adenine is always linked to thymine and guanine to cytosine (Figs 7.13 and 7.14). The number of hydrogen bonds formed between adenine and thymine is two, and the separation of the respective atoms in the hydrogen bond is about 0·3 nm (3 Å). The number of hydrogen bonds formed between cytosine and guanine is three. Guanine is never found hydrogen-bonded to adenine and if there is an adenine molecule in one chain then this demands a thymine molecule in the complementary chain. This gives the key to the replication of DNA in which the two strands of the helix may be unwound and separated and then a new strand is added to each of the original strands forming two daughter DNA molecules. These daughter DNA molecules have the same sequence of bases as in the original parent DNA molecule because the hydrogen bonding between the bases demands a particular sequence in the DNA molecule. The bases lie with their planes approximately at right angles to the helical axis, there being ten bases in each turn of the helix which are separated by an axial distance of 0·34 nm (3·4 Å) in each turn. Each turn has a height of 3·4 nm (34 Å) which gives the helix an exact tenfold screw axis. The sense of the screw is a right-handed one.

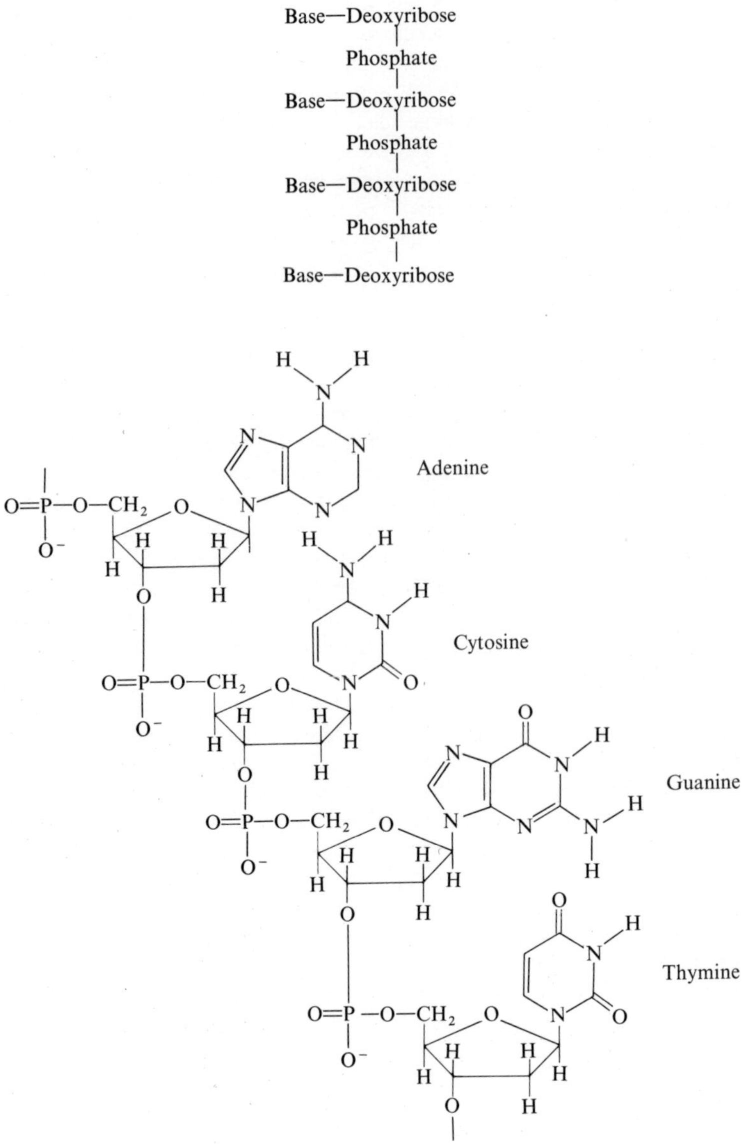

196

Fig. 7.12. The chemical structure of DNA

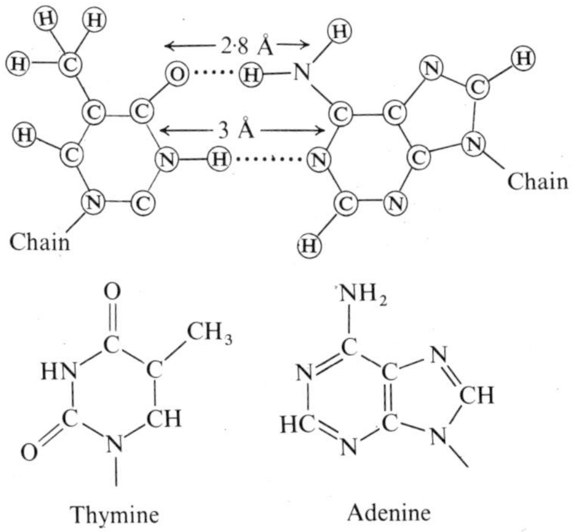

Fig. 7.13. The base pairing of thymine and adenine

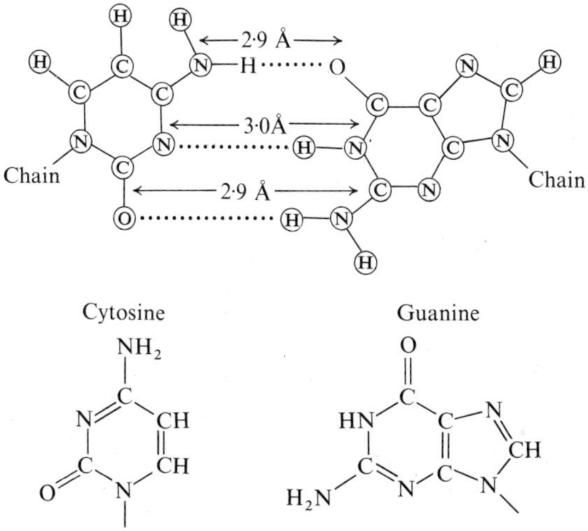

Fig. 7.14. The base pairing of cytosine and guanine

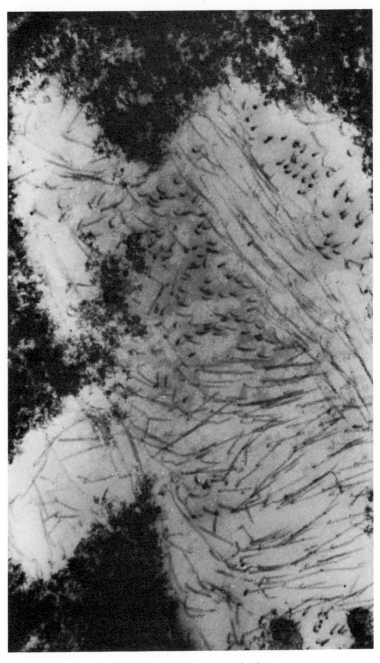

Plate 7.9. The tobacco mosaic virus

Viruses

The study of viruses plays an important part in the study of disease, not only of plants but also of human beings. The tobacco mosaic virus (Plate 7.9, see also Plates 2.11 and 2.12) is a particularly virulent virus which attacks tobacco leaves. Basically the (TMV) structure is a rod consisting of a threadlike molecule of RNA embedded in a helix formed of just over 2000 identical protein molecules. The sequence of the amino-acids is being

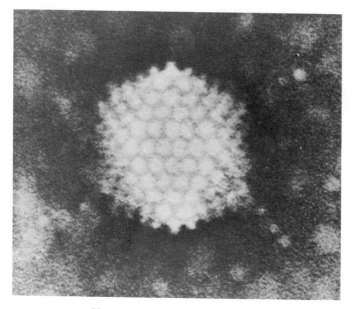

Plate 7.10. The adeno virus × 500 000

studied by chromotographic means and X-ray diffraction methods are now being applied to viruses of this kind. A study of the adeno virus (Plate 7.10) shows that the surface of the protein molecules has the form of an icosahedron. The magnification of this photograph is 500 000 times. Viruses act in a number of different ways but first the protein (the protection core) around the nucleic acid opens up and in some way the nucleic acid from the virus is injected into the cell of the species which is being attacked. The nucleic acid of the virus then governs the main genetic sequence. The T4 virus, which attacks bacteria, is an assembly of protein components consisting of a head, and a tail which constitutes a hypodermic syringe on a molecular level. The head of the syringe is a protein membrane which is shaped like a kind of prolate icosahedron with thirty facets and is filled with DNA. The head is attached to a tail which consists of a hollow core with a spike at the end away from the head. The spikes and fibres attached to the tail fix the virus to the bacterial cell wall, the viral DNA is then injected into the cell and the viral DNA is replicated rather than the original bacterial DNA. Synthesis of the virus then occurs inside the bacterial cell wall, which bursts releasing new virus particles.

Vitamin B_{12} and the porphyrins

Vitamin B_{12} was studied by Hodgkin and co-workers and is essential to normal metabolism, the deficiency of vitamin B_{12} being responsible for pernicious anaemia. The crystal consists of large molecules which are hexagonally close packed in layers. Each molecule consists of a cobalt atom which is surrounded by a corrin nucleus, which was new and completely unexpected at the time in organic chemistry, together with a cyanide ion. In the corrin nucleus the cobalt atom is surrounded by five five-membered rings, the nitrogen of which is attached by coordinate bonds to the cobalt. The carbon of the cyanide group is attached to the cobalt (Fig. 7.15).

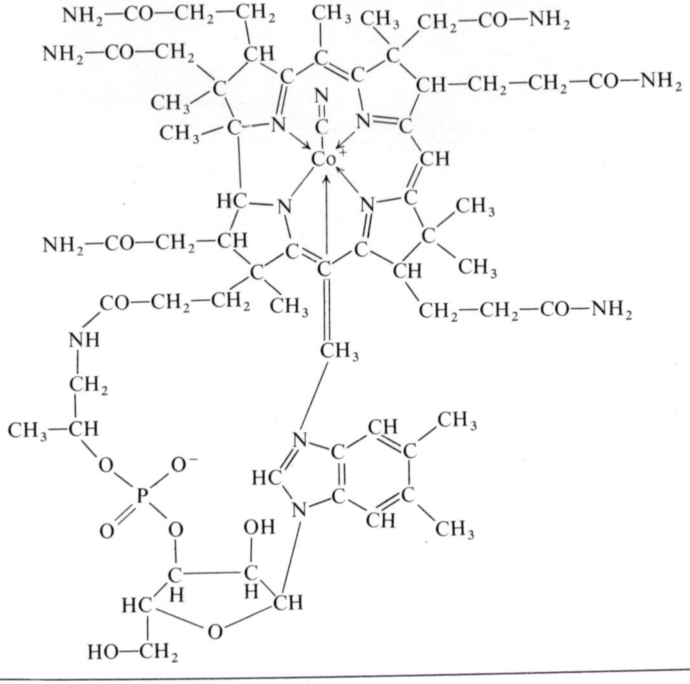

Fig. 7.15. The structure of vitamin B_{12}

Plate 7.11. A molecular model of wet crystals of vitamin B_{12}

The Science Museum has a model of wet crystals of vitamin B_{12}, a photograph of which is shown in Plate 7.11. One typical molecule of the vitamin B_{12} can be seen at the bottom right-hand side, with the cobalt atom being the large white ball in the centre of a network of atoms. To the right-hand side of this one can clearly see one of the five-membered rings of the corrin nucleus. Not much is known about the detailed chemistry of vitamin B_{12} but it appears that the vitamin itself is not biologically active, but it is easily converted when required into one or two of the natural vitamins including methylcobalamin which is thought to occur in liver. The corrin nucleus is distinct from, but related to the porphyrin nucleus; its structure in naturally occurring corrins was established by studying cobyric acid which crystallises in orange-red monoclinic crystals, which gave very good X-ray diffraction effects. From these crystals it was possible to derive a direct electron density map of the molecule in the crystal which showed the coordinates of the atoms except for the hydrogen atoms. The presence of the cobalt atom facilitated the determination of the face constants for nearly half of the observed X-ray diffraction reflections. The model was refined in the method already described, leading to a second electron density map from which the model shown could be built.

Penicillin

Chemical analysis could not distinguish between two possible structures for penicillin which was identified by X-ray analysis as the β lactam (Fig. 7.16). But how are these molecules arranged in space? One writes a chemical formula in the most simple manner possible so that a five-membered ring or six-membered ring is clearly shown in the plane of the paper. The chemical formula does not truly show the arrangements of the atoms in space and in sodium penicillin the molecule is not linear, but the molecules pack in the

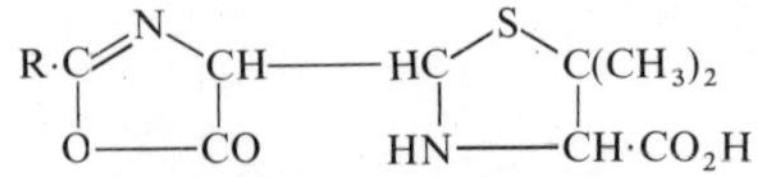

Oxazolone structure (false)

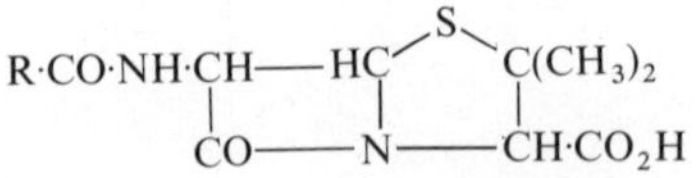

B. Lactam structure (true)

1. Two formulae from chemical evidence

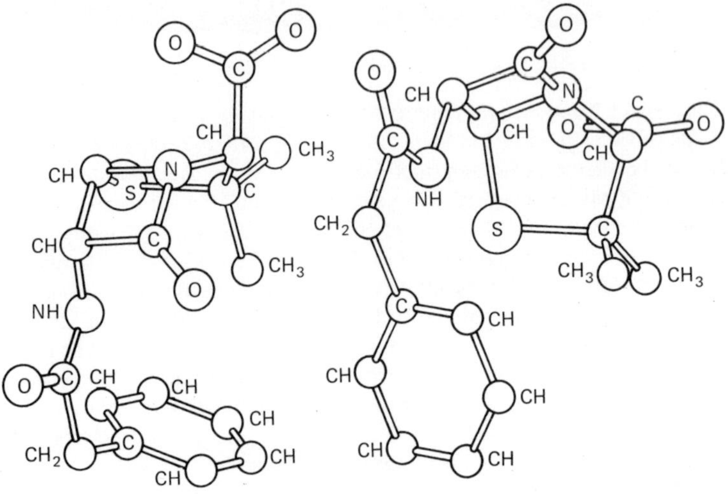

2. Arrangement of two molecules of the sodium salt in the lattice

Fig. 7.16. The structure of penicillin

manner shown. How is it then that a molecule such as penicillin can be used for the treatment of diseases? Recent experiments have shown that the molecule of penicillin interferes with normal bacterial growth by fitting into the space which would normally be occupied by amino-acids in the growing bacterial cell wall. Penicillin is a small molecule in which there is an inverse asymmetric centre but synthetic analogues of the opposite hand have no antibacterial activity and therefore the asymmetry of penicillin is important. The observation that asymmetry is important in antibacterial behaviour leads us back to the study of external structure which was studied in Chapter 2, because Pasteur, long ago, when he was led from his studies of crystal asymmetry and molecular asymmetry into the study of the treatment of diseases said, 'Who would now say what would be the future of germs if we could replace in these germs the immediate principles, albumen, cellulose, etc. by their inverse asymmetric principles?'

Glycerides

Glycerides are esters of glycerol and fatty acids, they are the principal constituents of fat and are widely found throughout the plant and animal kingdoms, being essential to living matter. The main function of glycerides is to store economically the biological energy needed for metabolism, yielding approximately twice as much energy per gramme as carbohydrates and protein. Glycerides are polymorphic and the different polymorphs may be obtained in the usual different ways. Why is it that the X-ray diffraction patterns indicate a short spacing of molecules along one axis, but long spacing along a different axis? The glycerides consist mainly of long chains and the short spacing corresponds to a cross section through the chain and the long spacing represents a section along the length of the chain. The carbon atoms have a tetrahedral valency distribution and the hydrocarbons chain is not straight but is a zigzag configuration. The molecules (Fig. 7.17) align themselves so that the axis along the chain in one molecule is parallel to the axis along the chain in another molecule. The total structure rather resembles a tuning fork and the prongs of the tuning fork fit in an opposite manner in the space lattice. How does polymorphism arise in such a structure? In the figure the chain of carbon atoms is in the plane of

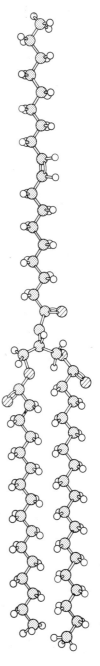

Fig. 7.17. Glycerides

the paper or in the plane of the tuning fork. One polymorphic structure arises if the chain AB is rotated through ninety degrees so that the chain of the carbon atoms is perpendicular to the plane of the tuning fork. Another polymorphic form arises when the zigzag carbon chains are not parallel.

Fibres

Before 1925 only three substances were recognised as being in fibrous forms. These were cellulose in cotton, flax and jute, keratin in wool, mohair, alpaca, etc., and collagen in real silk. Chemical analysis was useless because the sulphur content in two wool fibres, even within the same batch of wool, might vary by 200 per cent. X-ray diffraction quickly solved the problem of fibrous structure by noting that there was a short spacing and a long spacing, as in the glycerides, which meant that fibres were long molecules with the axes approximately parallel along the length of the fibre. Nature only uses two kinds of links for the formation of natural fibres: the link between glucose molecules in the cellulose fibres and the link which depends on the amino-acids, which is a characteristic structure of keratin and collagen. Chemists soon took advantage of this new knowledge and manufactured nylon by heating hexamethylene diamine and adipic acid (Fig. 7.18) to form a linear polymer, the chains of the linear polymer lying

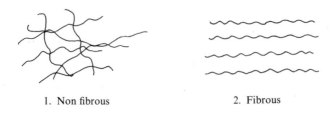

1. Non fibrous   2. Fibrous

Nylon

$H_2N.CH_2CH_2CH_2CH_2CH_2CH_2NH_2$ (hexamethylene diamine)
+
$HOOC.CH_2.CH_2.CH_2.CH_2.COOH$ (adipic acid)
↓
$...NH.CO.(CH_2)_4CONH.(CH_2)_6NH.CO(CH_2)_4CO...$

Terylene

$HO.CH_2.CH_2.OH$ (ethylene glycol)
+
$HOOC.C_6H_4.COOH$ (terephthalic acid)
↓
$...O(CH_2)_2OOC.C_6H_4O.(CH_2)_2OOC.C_6H_4...$

Fig. 7.18. Fibres

parallel to one another in nylon. Similarly, terylene was made by reacting ethylene glycol and terephthalic acid. Care must be taken, however, because spun terylene filaments are non crystalline and the molecules are

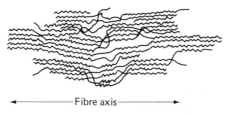

Fibre axis

1. Orientated crystallites (fibrous)

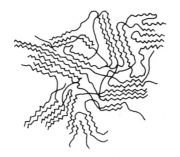

2. Unorientated crystallites; the material is resinous

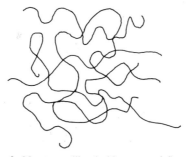

3. Non-crystalline (rubbery material)

Fig. 7.19. Relationship between fibres, resins and rubbers

unorientated. However, when the filaments are stretched an X-ray diffraction pattern may be obtained from the drawn fibre, showing that now the molecules are crystalline and well orientated. Even in well ordered fibres the regions of crystallinity do not usually exceed 50 per cent or 70 per cent. Within fibres (Fig. 7.19) there are areas of crystallinity but now the distribution of these areas of crystallinity gives a clue to the relationship between fibres,

Fig. 7.20. Decomposition of anthracene photo-oxide by X-rays

In practice the reaction depends on the existence of chains and is literally a chain reaction: the proportion of anthroquinone to anthrone depends on the lengths of the chains which go one way or the other; but the first stage in the reaction is the isolation of peroxide chains. This is easily proved by X-ray diffraction.

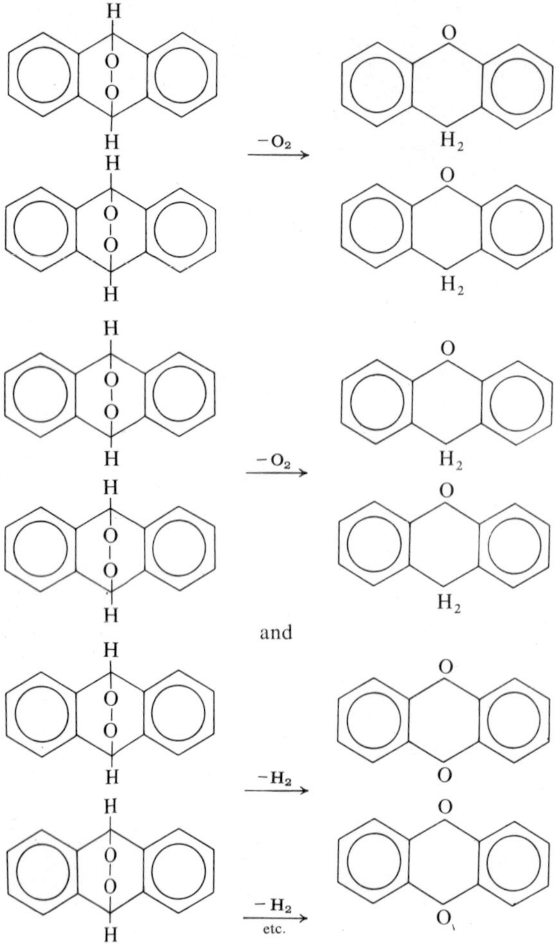

The ideal decomposition is

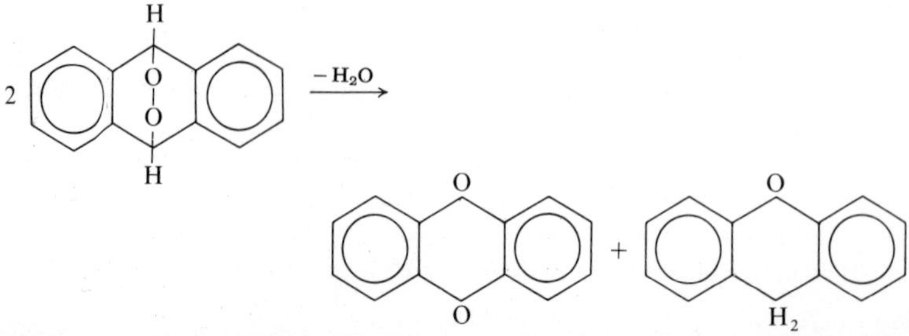

resins and rubbers. In resins there is less order of crystallinity than in fibres and in rubbers the molecules are distributed at random. In fibres, as we saw earlier in this chapter, the regions of order are maintained by hydrogen bonds.

Finally a word of warning. When one grows crystals of a substance X, one must be quite sure that the crystals grown are indeed X and not those of a new substance Y. Similarly it is important to know that the X-rays themselves do not cause any change in the composition. When crystals of anthracine photo-oxide were studied the resultant X-ray diffraction photographs could not be interpreted in terms of a structure of the photo-oxide, in fact what happens is that the substance decomposes to anthraquinone and anthrone (Fig. 7.20). An intuitive argument against any such rearrangement occurring might be that no hydrogen is seen bubbling off, but a comparison of the unit cell dimensions of the photo-oxide and the anthraquinone shows that one molecule of the photo-oxide occupies 0·254 nm^3 (2·54 Å^3) whereas one molecule of anthraquinone or anthrone only occupies 0·242 nm^3 (2·42 Å^3), leaving 5 per cent free space for the hydrogen atoms or water molecules to go into when slowly irradiated with X-rays. Indeed when quickly heated then the crystals of the photo-oxide swell and break into fibres, or even explode.

Questions

1 Describe how X-ray analysis has helped chemical study by reference to (a) vitamin B$_{12}$; (b) penicillin; (c) rubbers and fibres; (d) deoxyribonucleic acid.
2 How do the relationships between the structures of rubbers, resins and fibres serve to emphasise the importance of order in crystals?
3 Describe the importance of amino acids in (a) myoglobin; (b) haemoglobin; (c) insulin; (d) one virus.
4 Outline a normal X-ray analytical investigation on a substance on which chemical analysis only indicates the presence of carbon, hydrogen, oxygen and nitrogen. The substance has a high molecular weight, crystallises easily and there are indications that a heavy metal may be included into the crystal.
5 What are the limitations and advantages of X-ray diffraction?
6 Discuss the relative influence of ionic, covalent, hydrogen and van der Waal bonds on the structure. Draw examples from all chapters in this book.
7 Describe, with examples, how X-ray analysis has developed over the last fifty years.
8 Describe how a knowledge of the structure of simple molecules helps the understanding of the structure of larger molecules.

9 The following statements are meant to be provocative. Indicate to what extent you think the statements are true or false.

(a) Without hydrogen bonds we would all be dead.

(b) Myoglobin and haemoglobin are disordered molecules.

(c) Small molecules are always less important than large molecules.

(d) The cross sectional area of penicillin is 1 nm^2 (100 Å^2).

(e) Since X-ray analysis provides us with many of the fundamental structures we need, chemists are no longer necessary.

References

BRAGG, W. L., The Rutherford Memorial Lecture, 1960, *Proceedings of the Royal Society*, **262**, 1961, 145.

BRAGG, W. H., *The Structure of an Organic Crystal*, Fison Memorial Lecture, 1928, Longman.

BRAGG, W. L., 'Reminiscences of Fifty Years Research', *Proceedings of the Royal Institute*, **41**, 1966, 92.

CROWFOOT–HODGKIN, D., *Some Observations on Crystallography, Chemistry and Medicine*, The Harvey Lectures, Series 6, New York, Academic Press, page 205.

CROWFOOT-HODGKIN, D., 'Vitamin B_{12} and the porphyrins', *The Royal Society, Federation Proceedings*, **23**, 1964, 592.

DAVIES, D. R., 'X-ray Diffraction and the nucleic acids', *Chemistry* 40, 1967.

KENDREW, J. C., 'Myoglobin and the structure of proteins' *Science*, **139**, 1963, 1259.

LONSDALE, K., 'The defect mechanism of an organic single crystal chemical reaction (Decomposition of anthracene photo-oxide by X-rays)', International Conference on Electron Diffraction and Crystal Defects, Melbourne, 1965.

PERUTZ, M. F., *Proteins and Nucleic Acids: Structure and Function*, Eighth Weizmann Lecture Series, 1962, Elsevier Publishing Company.

PERUTZ, M. F., 'X-ray Analysis of Haemoglobin', Nobel Lecture, 11 December 1962.

TAYLOR, R. J., *The Chemistry of Glycerides*, Unilever Educational Booklets, Advanced Series No. 4.

TAYLOR, R. J., *The Chemistry of Proteins*, Unilever Educational Booklets, Advanced Series No. 3.

VALENTINE, R. C., 'Morphological and antigenic sub-units of viruses', *British Medical Bulletin*, **23**, 1967, 129.

VALENTINE, R. C., 'Portraits of proteins', *New Scientist*, 1 June 1967.

WATSON, J., *The Double Helix*, 1968, New York, Athenaeum Press.

WOOD, W. B., 'Building a bacterial virus', *Scientific American*, **217**, July 1967, 60.

Index